CW01572958

Prostasomes

Other titles in the Wenner-Gren International Series published by
Portland Press:

Symmetry *2000*
edited by I. Hargittai and T.C. Laurent
2002 ISBN 1 85578 149 2

Virtual University? Educational Environments of the Future
edited by H.J. van der Molen
2001 ISBN 1 85578 145 X

Advances in Strabismus Research: Basic and Clinical Aspects
edited by G. Lennerstrand and J.Ygge
2000 ISBN 1 85578 144 1

Plant Systematics for the 21st Century
edited by B. Nordenstam, G. El-Ghazaly and M. Kassas
2000 ISBN 1 85578 135 2

Lifelong Learning Policy and Research
edited by A. Tuijnman and T. Schuller
1999 ISBN 1 85578 134 4

The Impact of Electronic Publishing on the
Academic Community
edited by I. Butterworth
1998 ISBN 1 85578 122 0

WENNER-GREN INTERNATIONAL SERIES

VOLUME 81

Prostasomes

Proceedings from a symposium held at the
Wenner-Gren Centre, Stockholm, in June 2001

Edited by

G. Ronquist
B.O. Nilsson

PORTLAND PRESS
London

Published by Portland Press Ltd, 59 Portland Place, London W1B 1QW, U.K.
Tel: +44 (0)20 7580 5530; e-mail: editorial@portlandpress.com

ISBN 1 85578 151 4

British Library Cataloguing in Publication Data
A catalogue record for this book is available from the British Library

Typeset by Portland Press Ltd
Printed in Great Britain by Bell and Bain Ltd, Glasgow

Contents

This volume contains contributions presented at the First International Conference on Prostasomes, which was a Wenner-Gren Symposium held in Stockholm, Sweden, on 6th–9th June 2001.

The organizers wish to thank Professor Torvard Laurent for his efforts and advice as science secretary of the Wenner-Gren Foundations and Mrs Gun Lennerstrand for her excellent assistance during the planning and running of the conference. We also want to express our gratitude to the chairmen and speakers, who made the conference a fine forum for the exchange of new findings and novel ideas.

The conference presented a cross-section of the research now in progress in many laboratories on various functional and biochemical properties of prostasomes. Studies over the last few years have led to an increased understanding of the many peculiar aspects of prostasomes. For instance, prostasomes are surrounded by a membrane with a complex protein composition and a high cholesterol/phospholipid ratio, which endow them with many potent functions.

We hope that this volume will stimulate further work in this active field of research and provide new ideas for future experimentation.

Gunnar Ronquist
B. Ove Nilsson

Contributors

Anders Ahlander
Department of Medical Cell Biology, Unit of Anatomy, Biomedical Center, P.O. Box 571, SE-751 23 Uppsala, Sweden

Martin Albrecht
Department of Anatomy and Cell Biology, University of Marburg, Robert-Koch-Str. 6, D-35033 Marburg, Germany

Cinzia Allegrucci
Division of Animal Physiology, School of Biosciences, University of Nottingham, Sutton Bonington Campus, Loughborough LE12 5RD, U.K.

Giuseppe Arienti
Sezione di Biochimica, Dipartimento di Medicina Interna, Università di Perugia, Via del Giochetto, 06122 Perugia, Italy

Gerhard Aumüller
Department of Anatomy and Cell Biology, University of Marburg, Robert-Koch-Str. 6, D-35033 Marburg, Germany

Adil A. Babiker
Department of Medical Sciences, Clinical Chemistry, University Hospital, SE-751 85 Uppsala, Sweden

Anders Bjartell
Department of Urology, Malmö University Hospital, Malmö, Sweden

Julià Blanco
Fundació IRSI-Caixa, Hospital Universitari German Trias i Pujol, Badalona, Barcelona, Spain

Enrico Carlini
Sezione di Biochimica, Dipartimento di Medicina Interna, Università di Perugia, Via del Giochetto, 06122 Perugia, Italy

Lena Carlsson
Department of Medical Sciences, Clinical Chemistry, University Hospital, SE-751 85 Uppsala, Sweden

Cesare Castellini
Dipartimento di Scienze Zootecniche, Università di Perugia, Perugia, Italy

Francisco Ciruela
Department Bioquímica i Biologia Molecular, Facultat de Química, Martí i Franquès 1, 08028 Barcelona, Spain

Ingrid De Meester
Department of Pharmaceutical Sciences, Laboratory of Medical Biochemistry, Universiteitsplein 1, B-2610 Antwerp, Belgium

Lars Egevad
Department of Pathology and Cytology, Karolinska Hospital, Stockholm, Sweden

Elena Faccenda
Medical Research Council Human Reproductive Science Unit, University of Edinburgh Centre for Reproductive Biology, 37 Chalmers Street, Edinburgh EH3 9ET, U.K.

Rafael Franco
Department Bioquímica i Biologia Molecular, Facultat de Química, Martí i Franquès 1, 08028 Barcelona, Spain

Gilles Frenette
Unité d'Ontogénie-Reproduction, Centre de Recherche, Centre Hospitalier de l'Université Laval, 2705 Blvd. Laurier, Ste-Foy, PQ, Canada, G1V 4G2

Ara G. Hovanessian
Unité de Virologie et d'Immunologie Cellulaire, UA CNRS 1157, Institut Pasteur, 28 rue du Dr. Roux, 75724 Paris Cedex 15, France

Rodney W. Kelly
Medical Research Council Human Reproductive Science Unit, University of Edinburgh Centre for Reproductive Biology, 37 Chalmers Street, Edinburgh EH3 9ET, U.K.

Masaya Kitamura
Department of Urology, Osaka National Hospital, 2-1-14 Hoenzaka, Chuo-ku, Osaka 540-0006, Japan

Anne-Marie Lambeir
Department of Pharmaceutical Sciences, Laboratory of Medical Biochemistry, Universiteitsplein 1, B-2610 Antwerp, Belgium

Anders Larsson
Department of Medical Sciences, Clinical Chemistry, University Hospital, SE-751 85 Uppsala, Sweden

Christine Légaré
Unité d'Ontogénie-Reproduction, Centre de Recherche, Centre Hospitalier de l'Université Laval, 2705 Blvd. Laurier, Ste-Foy, PQ, Canada, G1V 4G2

Lavinia Liguori
Dipartimento di Scienze Biochimiche e Biotecnologie Molecolari, Sezione di Biochimica Cellulare, Via del Giochetto, 06123 Perugia, Italy

Carmen Lluis
Department Bioquimíca i Biologia Molecular, Facultat de Química, Martí i Franqués 1, 08028 Barcelona, Spain

Kiyomi Matsumiya
Department of Urology, Osaka University Graduate School of Medicine, Suita, Japan

Alba Minelli
Dipartimento di Scienze Biochimiche e Biotecnologie Molecolari, Sezione di Biochimica Cellulare, Via del Giochetto, 06123 Perugia, Italy

Bo Nilsson
Department of Oncology, Radiology and Clinical Immunology, University of Uppsala, Uppsala, Sweden

B. Ove Nilsson
Department of Medical Cell Biology, Unit of Anatomy, Biomedical Center, P.O. Box 571, SE-751 23 Uppsala, Sweden

Ulf R. Nilsson
Department of Oncology, Radiology and Clinical Immunology, University of Uppsala, Uppsala, Sweden

Bo-Johan Norlén
Department of Urology, University Hospital, Uppsala, Sweden

Toshitugu Oka
Department of Urology, Osaka National Hospital, 2-1-14 Hoenzaka, Chuo-ku, Osaka 540-0006, Japan

Akihiko Okuyama
Department of Urology, Osaka University Graduate School of Medicine, Suita, Japan

Carlo Alberto Palmerini
Dipartimento di Scienze Biochimiche e Biotecnologie Molecolari, Università di Perugia, Perugia, Italy

Gunnar Ronquist
Department of Medical Sciences, Clinical Chemistry, University Hospital, SE-751 85 Uppsala, Sweden

Fabrice Saez
Laboratoire de Biologie de la Reproduction, Université d'Auvergne, 28, Place Henri Dunant, 63000 Clermont-Ferrand, France

Göran Sahlén
Department of Urology, University Hospital, Uppsala, Sweden

Simon Scharpé
Department of Pharmaceutical Sciences, Laboratory of Medical Biochemistry,
Universiteitsplein 1, B-2610 Antwerp, Belgium

Jürgen Seitz
Department of Anatomy and Cell Biology, University of Marburg, Robert-Koch-Str. 6,
D-35033 Marburg, Germany

Tsukasa Seya
Department of Immunology, Osaka Medical Center for Cancer and Cardiovascular
Diseases, Osaka, Japan

Grzegorz Skibinski
Department of Clinical Biochemistry, Queen's University, Grosvenor Road,
Belfast BT12 6BJ, Northern Ireland, U.K.

Jian Song
Department of Anatomy and Cell Biology, University of Marburg, Robert-Koch-Str. 6,
D-35033 Marburg, Germany

Mats Stridsberg
Department of Medical Sciences, Clinical Chemistry, University Hospital, SE-751 85
Uppsala, Sweden

Robert Sullivan
Unité d'Ontogénie-Reproduction, Centre de Recherche, Centre Hospitalier de
l'Université Laval, 2705 Blvd. Laurier, Ste-Foy, PQ, Canada, G1V 4G2

Akira Tsujimura
Department of Urology, Osaka University Graduate School of Medicine, Suita, Japan

Agustín Valenzuela
Unité de Virologie et d'Immunologie Cellulaire, UA CNRS 1157, Institut Pasteur,
28 rue du Dr. Roux, 75724 Paris Cedex 15, France

Beate Wilhelm
Department of Anatomy and Cell Biology, University of Marburg, Robert-Koch-Str. 6,
D-35033 Marburg, Germany

Michael J. Wilson
Research Service, VA Medical Center, One Veterans Drive, Minneapolis, MN 55417,
U.S.A.

Abbreviations

A23187-AR	A23187-stimulated acrosome reaction
ACE	angiotensin-converting enzyme
ACT	α_1-anti-chymotrypsin
ADA	adenosine deaminase
ADAbp	ADA binding protein
APUD	amine precursor uptake and decarboxylation
AR	acrosome reaction
ASA	anti-sperm antibodies
$[Ca^{2+}]_i$	intracellular Ca^{2+} concentration
CD	cluster of differentiation
CgA	chromogranin A
CgB	chromogranin B
Con A	concanavalin A
DAF	decay-accelerating factor
DPP IV	dipeptidyl peptidase IV
ecto-ADA	cell-surface ADA
f-C3(MA)	methylamine-treated fluorescently labelled C3
fMLP	formyl-Met-Leu-Phe
GPI	glycosylphosphatidylinositol
ICSI	intra-cytoplasmic sperm injection
IL	interleukin
IUI	intra-uterine insemination
IVF	*in vitro* fertilization
LIF	leukaemia inhibitory factor
MAPK	mitogen-activated protein kinase
MCLA	2-methyl-6-(*p*-methoxyphenyl)-3,7-dihydroimidazo-[1,2-*a*]pyrazin-3-one
MCP	membrane cofactor protein
MV	measles virus
NEP	neutral endopeptidase
NHS	normal human serum
NK	natural killer
NP-40	Nonidet P-40
NRR	non-return rate
PAP	prostatic acid phosphatase
PCI	protein C inhibitor
Pg2	plasminogen type 2
PGE	prostaglandin E
PIPLC	phosphatidylinositol-specific phospholipase C
PKA	cAMP-dependent protein kinase
PKC	protein kinase C
PKI	protein kinase inhibitor protein
PLP	prostasome-like particle
PMN	polymorphonuclear neutrophils
PNH	paroxysmal nocturnal haemoglobinuria
PSA	prostate-specific antigen
R_{18}	octadecylrhodamine

RE	rabbit erythrocyte
ROS	reactive oxygen species
SCID	severe combined immunodeficiency syndrome
SgII	secretogranin II
ssCD46	seminal plasma soluble CD46
TF	tissue factor
TGFβ	transforming growth factor β

Prostasomes

Gunnar Ronquist*[1], Lena Carlsson*, Anders Larsson* and B. Ove Nilsson†
*Department of Medical Sciences, Clinical Chemistry, University Hospital, SE-751 85
Uppsala, Sweden, and †Department of Medical Cell Biology, Unit of Anatomy,
Biomedical Center, P.O. Box 571, SE-751 23 Uppsala, Sweden

Background

ATPases are structure-bound enzymes that hydrolyse the terminal phosphate
from ATP, thus giving rise to ADP and P_i. As is well known, this reaction is
exergonic with a $\Delta G'$ value of about -29.3 kJ/mol (-7 kcal/mol). These ATPases
are organized in membranous structures and assembled in phospholipoprotein
complexes. A high ATPase activity was found to be linked to membranous
structures of human prostatic fluid and seminal plasma [1–3]. This ATPase was
Mg^{2+}- and/or Ca^{2+}-dependent, thus differing from the Na^+/K^+-ATPase and
exhibiting no dependence on Na^+ and K^+, and ouabain was inefficient as an
inhibitor [2,3]. Deoxycholate, Triton X-100, dodecanylsulphate and oleate were
all potent inhibitors of the ATPase [2–4], underlining its phospholipoprotein
nature and membrane association. This ATPase activity was recovered in a pellet
after preparative ultracentrifugation of prostatic fluid and seminal plasma. The
pelleted material was retrieved in a main fraction at a density of 1.03 upon gradient
centrifugation in a self-generating colloidal silica solution retaining the ATPase
activity [2,3]. The ATPase activity was membrane-linked to structures named
prostasomes [5,6].

Origin of prostasomes in humans

When comparing prostatic fluid and seminal plasma from the same individuals
with regard to ATPase activity and fructose content the following was found. In
prostatic fluid a low fructose content (a seminal vesicle secretory product) was
accompanied by a significantly higher ATPase activity. This relationship was
reversed in seminal plasma, which meant that the ATPase activity was 'diluted'
while fructose was enriched by the contribution of the seminal vesicles [2,3].
Hence, this finding gave support to the view that the ATPase activity originated
solely from the prostate gland. Also, investigating split ejaculate fractions revealed
a typical pattern. The highest ATPase activity and highest frequency of organellar
structures (prostasomes) were found in the first two fractions, representing the
prostatic contribution, whereas the lowest ATPase activity and prostasome
occurrence was typical of the last two fractions, representative of the seminal
vesicles' contribution [2,3]. A strong correlation was obtained between total
ATPase activity in different split ejaculate fractions and each of the bivalent

[1]To whom correspondence should be addressed (e-mail gunnar.ronquist@clm.uas.lul.se).

cations, Zn^{2+}, Ca^{2+} and Mg^{2+} [7]. These bivalent cations are secreted only by the prostate gland and therefore this high correlation supported the prostatic origin of the ATPase complex.

Direct studies were also carried out on human prostatic tissue from four men aged 59–68 years who were undergoing radical cystectomy, including total prostatectomy, because of bladder cancer. Immediately after the operation the prostate gland was dissected free from the bladder, seminal vesicles and adhering tissues. The prostate was then dissected further into three paired lobes, as described in [8]. The material was placed immediately in a moist chamber at 0°C and kept in a solution of ice and water to maintain the low temperature. Within a few hours, secretory fluid was squeezed gently from the ductules of the cut prostatic sections and collected in small plastic tubes. In three patients the prostatic dissection disclosed well-defined peri-urethral hyperplastic nodules, the largest with a diameter of 20 mm. The nodules were removed from the prostatic tissue without rupture of the surrounding capsule. Secretion was obtained from solitary nodules by gentle squeezing. No contamination with prostatic tissue material was possible, as the nodules were prepared separately. The squeezed fluid from prostatic tissue and hyperplastic nodules was diluted 1:3 (v/v) with 0.15 M NaCl solution and centrifuged at 2000 g for 10 min. The residual pellet, containing cells and cell debris, was discarded and the supernatant was used for biochemical analyses. In all cases the Mg^{2+}- and Ca^{2+}-dependent ATPase activity could be recorded and was present at a level higher than that expected for mixed seminal plasma (which also contains secretions from seminal vesicles, testes and epididymes). The hyperplastic nodules essentially contained the same substances as the prostatic tissue (Table 1). The values for acid phosphatase activity, calcium, magnesium and zinc were as expected for prostatic fluid (Table 1). The concentrations of these substances varied to some extent within different prostatic lobes, but without statistically significant differences (Table 2). A contribution of prostasomal ATPase from spermatozoa was ruled out for the following reason: 13 men were investigated before and after vasectomy with regard to the prostasomal ATPase activity in their semen and no change was noted in the mean ATPase activities. Instead, since the ATPase activity was expressed per semen volume unit, there was a 25% increase in this ATPase activity, explained by the reduction in volume of the ejaculate due to the vasectomy [9].

The Mg^{2+}- and Ca^{2+}-dependent ATPase associated with the prostasome membrane was most probably of the P type. This means that it undergoes intermediate phosphorylation during catalysis, and is inhibited by vanadate. Indeed, low concentrations of vanadate ($<100 \, \mu M$) were found to be inhibitory [10]. P-type ATPases include the Na^+/K^+-ATPase and Ca^{2+}-ATPase of the plasma membrane and sarcoplasmic reticulum ('transport ATPases'). The finding that low concentrations of calmidazolium and oleate were inhibitory to the prostasomal ATPase system [4] was in line with expectations for a P-type ATPase.

Table1

	ATPase (μmol· min^{-1}·g of protein^{-1})	AP (μkat·g of protein^{-1})	Cation (μmol·g of protein^{-1})		
			Ca^{2+}	Mg^{2+}	Zn^{2+}
Prostatic gland					
Mean	54.5	235	171	44.4	21.0
S.D.	41.3	197	111	17.1	12.0
Range	8.2–170	59.0–684	72.2–396	24.8–68.0	7.9–47.0
Hyperplastic nodules					
Mean	50.1	129.3	135	54.2	21.6
S.D.	16.5	53.5	48.7	29.6	5.3
Range	38.8–69.0	91.0–190	98.6–190	34.8–88.3	16.3–26.9

ATPase, acid phosphatase (AP), Ca^{2+}, Mg^{2+} and Zn^{2+} in fluid from the prostate glands of four men and adenomatous tissue from three of the same glands

Data are from Stegmayr, B., Busch, C., Fritjofsson, Å. and Ronquist, G. (1985) Upsala J. Med. Sci. **90**, 139–145.

Table 2

Lobe	ATPase (μmol· min⁻¹·g of protein⁻¹)	AP (μkat·g of protein⁻¹)	Cation (μmol·g of protein⁻¹)		
			Ca^{2+}	Mg^{2+}	Zn^{2+}
Dorsal	40.9 (24.3)	258.9 (212.9)	178.0 (101.8)	44.1 (20.8)	23.2 (12.6)
Lateral	75.7 (64.1)	298.0 (269.0)	144.0 (114.0)	40.1 (11.1)	21.9 (17.5)
Medial	42.0 (15.1)	149.0 (95.2)	192.0 (141.0)	49.1 (21.4)	17.8 (6.2)

Values for prostatic fluid according to lobe of origin

*Lobes from the glands of four men were tested. Means (and S.D.) are shown. Data are from Stegmayr, B., Busch, C., Fritjofsson, Å. and Ronquist, G. (1985) Upsala J. Med. Sci. **90**, 139–145.*

AP, acid phosphatase.

The effect of concanavalin A (Con A) on the prostasomal ATPase system

In itself, Con A had a negligible effect on the prostasomal ATPase system [1]. However, when prostasomes were preincubated with a detergent (e.g. deoxycholate) the subsequent assay for ATPase showed a sharp inhibition of ATPase activity. If this first preincubation was followed by another preincubation with low concentrations of Con A, then the detergent-induced inhibition of the Mg^{2+}- and Ca^{2+}-dependent ATPase system was abrogated [1]. Also, if the sequence of additions was reversed prior to the ATPase assay, i.e. Con A was added before the detergent, there was no inhibitory action at all by the detergent [1]. Con A is a lectin that binds to saccharides containing α-D-mannopyranosyl or α-D-glucopyranosyl residues [11]. The effect of Con A on the prostasomal ATPase system was specific, because α-methyl-D-mannoside nullified it [1]. How can we explain the effect of Con A, which indeed emphasizes the membranous context of the prostasomal ATPase system? At least two explanations are available for Con A's mediation of increased ATPase expression in the membranous material after inactivation with detergent. One would envision the binding of Con A to a surface carbohydrate inhibitor of the prostasomal ATPase. This means that the inhibitor is normally not in direct contact with the membrane-bound ATPase system due to the presence of neighbouring phospholipids. Accordingly, inhibition is only accomplished if the phospholipids are removed by a detergent, and this inhibition can be abolished by Con A binding to the carbohydrate inhibitor.

The other explanation postulates Con A binding to the ATPase enzyme system, leading to stimulation. Con A would in such a case be a positive allosteric effector. This would be in agreement with results showing that membrane-bound enzymes such as ATPases display co-operativity towards various allosteric effectors [12,13].

Androgen dependency of prostasome occurrence in seminal plasma

There are reasons to believe that testosterone is decisive for the appearance of prostasomes in seminal fluid. First, a pronounced deficiency of Mg^{2+}/Ca^{2+}-ATPase and of prostasomes was demonstrated on repeated occasions in seminal plasma from a patient with infertility problems and with a low serum testosterone level [14]. Secondly, 22 oligo-zoospermic men with reduced testosterone levels had a mean 30% reduction in their prostasomal ATPase activity compared with 30 normo-zoospermic men [9]. Thirdly, another patient with a well-differentiated carcinoma of the prostate demonstrated an 85% drop in prostasomal ATPase activity after 14 days of anti-androgenic treatment, compared with his pretreatment value [15].

The prostasomal membrane architecture

The prostasomal membrane is composite in its construction and this is valid for both the lipid part and the protein part. Lipid analysis of prostasomes revealed a striking quantitative domination of cholesterol over the phospholipids, the molar proportions of cholesterol/sphingomyelin/glycerophospholipids being 4:1:1. Hence, the cholesterol/phospholipid ratio, being 2.0 [16], was very high in comparison with most other plasma membranes, including that of the human spermatozoon (0.83) [17]. This high ratio was consonant with a high percentage of sphingomyelin in prostasomes, giving the membrane a high degree of molecular ordering, as measured by ESR [16].

Prostasomes subjected to SDS/PAGE yielded a complex pattern [18,19]. Typically, 18 main bands can be discerned, some of which have been identified. The high molecular-mass bands represent CD (cluster of differentiation) molecules and are CD13 (aminopeptidase; 150 kDa), CD26 (dipeptidyl peptidase IV, 120 kDa) and CD10 (enkephalinase; 100 kDa). The aminopeptidase, a zinc-dependent proteolytic enzyme, was found to be a useful marker of prostasomes [10,20,21]. However, it is not involved in any ATP-driven protease activity in prostasomes [10]. Dipeptidyl peptidase IV activity was found to be extremely high in prostasomes [22]. The dipeptidyl peptidase IV antigen (CD26) was identified on prostasomes using a monoclonal antibody [23], and the presence of both the enzyme activity and the antigen was established in prostatic secretions, but was absent from seminal vesicle secretions [24]. This finding gives further support to the view that prostasomes are exclusive secretory products of the prostate with no additional supply from the seminal vesicles. Dipeptidyl peptidase IV is a membrane-associated proteinase that cleaves off dipeptides from peptide chains that consist of three or more amino acid residues. Its primary specificity is for proline at the penultimate position in the N-terminus [25].

As a serine protease with unique specificity, dipeptidyl peptidase IV participates in peptide metabolism during the intestinal digestion and renal transport of polypeptides. In the haematopoietic system, dipeptidyl peptidase IV is involved in the activation of T-lymphocytes and the regulation of DNA synthesis, cell proliferation and the production of cytokines (reviewed by Yaron and Naider [25] and Fleischer [26]). The enzyme may also play a role in HIV infection and apoptosis [27]. The HIV-1 Tat (trans-activating) protein binds the dipeptidyl peptidase IV enzyme and inhibits its activity [28], and the HIV-1 envelope glycoprotein gp120 also interferes with it [29]. Therefore, prostasomes may well bind HIV viruses via prostasome membrane-bound dipeptidyl peptidase IV. It should be mentioned in this context that prostasomes also contain complement inhibitors such as CD46 [30], CD55 and CD59 [31]. Complement inhibitors may be present in semen to protect the spermatozoa, but they might protect the pathogens as well [32]. The interaction between complement inhibitors and viruses has been investigated in *in vitro* experimental systems. The HIV virus, after incubation with CD55 and CD59, acquires these inhibitors in its membrane, and this leads to an increased resistance against attack by complement [33]. Since these complement inhibitors do exist at the prostasome membrane surface, the same working mechanism may be valid in the presence of prostasomes. Accordingly, it may be possible for us to discern a new principle here, by which prostasomes convey an advantage to the HIV virus in terms of survival in human semen.

The 100 kDa protein from human prostasomes was recently identified as neutral endopeptidase (NEP, or enkephalinase; CD10) using N-terminal amino acid sequencing and mass fingerprint analysis [34]. NEP was initially characterized as a zinc metalloendopeptidase with a specificity similar to a group of microbial enzymes that includes thermolysin. NEP was identified as the synaptic peptidase (enkephalinase) involved in terminating the actions of the opioid peptides, the enkephalins [35,36]. Among its potential substrates are the opioid peptides, tachykinins such as substance P, the natriuretic peptide family, bombesin-like peptides and chemotactic peptides. A small and variable number of neuroendocrine cells of the epithelial lining of prostatic tissue seemed to contain bombesin-like immunoreactivity [37]. Molecular cloning of NEP revealed it to be a type-II integral membrane protein consisting of a short N-terminal cytoplasmic domain of 27 amino acids, a transmembrane region of 22 hydrophobic residues and a large extracellular domain of about 700 residues that contains the enzyme's active site [38]. This means that NEP is one of the family of membrane-anchored ectoenzymes. It was shown subsequently [39] that NEP is identical to CD10, a tumour-associated cluster-differentiation antigen expressed on the surface of neutrophils and certain lymphoid progenitors, and also known as the common acute lymphoblastic leukaemia antigen.

Sperm–prostasome interaction

The pivotal role of prostasomes in the fertilization process implicates a sperm–prostasome interaction. Indeed, prostasomes can adhere to and, at least to some extent, fuse with sperm cells, as shown by free-zone electrophoresis and electron microscopy [40], octadecyl-rhodamine fluorescence self-quenching [41] and immunofluorescence staining and confocal microscopy [42]. Both spermatozoa and prostasomes display a net negative surface charge when analysed by free-zone electrophoresis and, on comparison, prostasomes are slightly less negative than spermatozoa [40]. Accordingly, prostasomes, at first ahead of the spermatozoa on the electrophoretic gels, were reached after a 20–25 min run by the spermatozoa as a result of their faster mobility, and the two, originally distinct, fractions joined into one common fraction, which did not dissociate [40]. If there was no interaction then the spermatozoa should have passed the prostasome fraction and formed its own fraction ahead of the prostasomes on the gel. The observed interaction took place regardless of prior neuraminidase treatment of either prostasomes alone or both prostasomes and spermatozoa. This, together with the finding that Con A-pretreatment of prostasomes and spermatozoa did not interfere with the interaction, made it less probable that carbohydrates were involved in the bonding. It was concluded that the bonding was most probably hydrophobic in nature [40]. This conclusion was also corroborated by the fact that an electron-microscopic demonstration of a prostasome–sperm interaction was only possible when embedding of the fused fraction after free-zone electrophoresis was carried out in a hydrophilic resin [40].

Conclusions

Prostasomes are prostate-derived sub-micron-sized organelles occurring in human semen. They have several biological activities but their physiological function is still not fully known. The membrane surrounding the prostasomes exhibits a very high cholesterol/phospholipid ratio, giving rise to a high degree of molecular ordering. It is assumed that prostasomes convey their various abilities to sperm cells by coating them, and as a consequence the sperm cells are better prepared for their ultimate goal of reaching and fertilizing the ovum.

References
1   Ronquist, G. and Hedström, M. (1977) Biochim. Biophys. Acta **483**, 483–486
2   Ronquist, G., Brody, I., Gottfries, A. and Stegmayr, B. (1978) Andrologia **10**, 261–272
3   Ronquist, G., Brody, I., Gottfries, A. and Stegmayr, B. (1978) Andrologia **10**, 427–433
4   Ronquist, G. (1987) Eur. J. Clin. Invest. **17**, 231–236
5   Ronquist, G. and Brody, I. (1982) First European Congress on Cell Biology, July 18–23, Paris, Abstract 701, ECCB, Paris
6   Stegmayr, B. and Ronquist, G. (1982) Urol. Res. **10**, 253–257
7   Stegmayr, B., Berggren, P.-O., Ronquist, G. and Hellman, B. (1982) Scand. J. Urol. Nephrol. **16**, 199–203
8   Salander, H., Johansson, S. and Tisell, L.E. (1981) Invest. Urol. **18**, 479–483
9   Stegmayr, B., Gottfries, A., Ronquist, G. and Brody, I. (1980) Scand. J. Urol. Nephrol. **14**, 129–134
10   Ronquist, G. (1988) Urol. Int. **43**, 334–340
11   Lis, H. and Sharon, N. (1973) Annu. Rev. Biochem. **42**, 541–574
12   Goldemberg, A.L., Farias, R.N. and Trucco, R.E. (1972) J. Biol. Chem. **247**, 4299–4309
13   Sineriz, F., Farias, R.N. and Trucco, R.E. (1973) FEBS Lett. **32**, 30–32
14   Brody, I., Ronquist, G., Gottfries, A. and Stegmayr, B. (1981) Scand. J. Urol. Nephrol. **15**, 85–90
15   Ronquist, G. and Stegmayr, B. (1984) Urol. Res. **12**, 243–247
16   Arvidson, G., Ronquist, G., Wikander, G. and Öjteg, A.-C. (1989) Biochim. Biophys. Acta **984**, 167–173
17   Mack, S.R., Everingham, J. and Zaneveld, L.J.D. (1986) J. Exp. Zool. **240**, 127–136
18   Ronquist, G. and Brody, I. (1985) Biochim. Biophys. Acta **822**, 203–218
19   Lindahl, M., Tagesson, C. and Ronquist, G. (1987) Urol. Int. **42**, 385–389
20   Laurell, C.B., Weiber, H., Ohlsson, K. and Rannevik, G. (1982) Clin. Chim. Acta **126**, 161–170
21   Ronquist, G., Frithz, G. and Jansson, Å. (1988) Urol. Int. **43**, 133–138
22   Vanhoof, G., De Meester, I., van Sande, M., Scharpe, S. and Yaron, A. (1992) Eur. J. Clin. Chem. Clin. Biochem. **30**, 333–338
23   Schrimpf, S.P., Hellman, U., Carlsson, L., Larsson, A., Ronquist, G. and Nilsson, B.O. (1999) Prostate **38**, 35–39
24   Wilson, M.J., Ruhland, A.R., Pryor, J.L., Ercole, C., Sinha, A.A., Hensleigh, H., Kaye, K.W., Dawkins, H.J.S., Wasserman, N.F., Reddy, P. and Ahmed, K. (1998) J. Urol. **160**, 1905–1909
25   Yaron, A. and Naider, F. (1993) Crit. Rev. Biochem. Mol. Biol. **28**, 31–81
26   Fleischer, B. (1994) Immunol. Today **15**, 180–184
27   Schlossman, S.F. (1994) Proc. Natl. Acad. Sci. U.S.A. **91**, 9960–9964
28   Subramanyam, M., Gutheil, W.G., Bachovshin, W.W. and Huber, B.T. (1993) J. Immunol. **150**, 2544–2553
29   Valenzuela, A., Blanco, J., Callebaut, C., Jacotot, E., Lluis, C., Hovanessian, A.G. and Franco, R. (1997) J. Immunol. **158**, 3721–3729
30   Simpson, K.L. and Holmes, C.H. (1994) J. Reprod. Fertil. **102**, 419–424
31   Rooney, I.A., Atkinson, J.P., Krul, E.S., Schonfled, G., Polakoski, K., Saffitz, J.E. and Morgan, B.P. (1993) J. Exp. Med. **177**, 1409–1420
32   Rooney, I.A., Heuser, J.E. and Atkinson, J.P. (1996) J. Clin. Invest. **97**, 1675–1686
33   Saifuddin, M., Parker, C.J., Peeples, M.E., Gorny, M.K., Zolla-Pazner, S., Ghassemi, M., Rooney, I.A., Atkinson, J.P. and Spear, G.T. (1995) J. Exp. Med. **182**, 501–509
34   Renneberg, H., Albrecht, M., Kurek, R., Krause, E., Lottspeich, F., Aumüller, G. and Wilhelm, B. (2001) Prostate **46**, 173–183

35   Relton, J.M., Gee, N.S., Matsas, R., Turner, A.J. and Kenny, A.J. (1983) Biochem. J. **215**, 519–523

36   Matsas, R., Fulcher, I.S., Kenny, A.J. and Turner, A.J. (1983) Proc. Natl. Acad. Sci. U.S.A. **80**, 3111–3115

37   diSant'Agnese, P.A. (1986) Arch. Pathol. Lab. Med. **110**, 412–415

38   Devault, A., Lazure, C., Nault, C., Le Moual, H. Seidah, N.G., Chrétien, M., Kahn, P., Powell, J., Mallet, J., Beaumont, A. et al. (1987) EMBO J. **6**, 1317–1322

39   LeTarte, M., Vera, S., Tran, R., Addis, J.B.L., Onizuka, R.J., Quackenbush, E.J., Jongeneel, C.V. and McInnes, R.R. (1988) J. Exp. Med. **168**, 1247–1253

40   Ronquist, G., Nilsson, B.O. and Hjertén, S. (1990) Arch. Androl. **24**, 147–157

41   Carlini, E., Palmerini, C.A., Cosmi, E.V. and Arienti, G. (1997) Arch. Biochem. Biophys. **343**, 6–12

42   Minelli, A., Allegrucci, C., Mezzasoma, I., Ronquist, G., Lluis, C. and Franco, R. (1999) Biol. Reprod. **61**, 802–808

2

Visualizing prostasomes: ultrastructure of the secretory machinery in prostate epithelium from benign prostate hyperplasia and prostate adenocarcinoma

Göran Sahlén*, Lars Egevad†, Anders Ahlander‡, Bo-Johan Norlén*, Gunnar Ronquist§ and B. Ove Nilsson‡[1]

*Department of Urology, University Hospital, Uppsala, Sweden, †Department of Pathology and Cytology, Karolinska Hospital, Stockholm, Sweden, ‡Department of Medical Cell Biology, Unit of Anatomy, Biomedical Center, P.O. Box 571, SE-751 23 Uppsala, Sweden, and §Department of Medical Sciences, Clinical Chemistry, University Hospital, SE-751 85 Uppsala, Sweden

Introduction

Under the electron microscope, the secretory activity in the prostate epithelium can be observed as apical vesicles (storage vesicles) containing a dense substance and scattered prostasomes, that is, small dark granules and small vesicles [1–6]. When the storage vesicles have released their contents into the gland ducts, these will be filled with a fuzzy substance and aggregates of prostasomes.

The creation of the storage vesicles and prostasomes is to a large extent performed by the Golgi apparatus, the organelle that packages and delivers secretory products in a cell. During transport from the Golgi apparatus to the apical part of the prostate cell, the storage vesicles and their contents undergo a maturation process before being expelled into the glandular lumen. To learn about the structural counterparts of these various processes, we have examined the secretory machinery in prostate cells by transmission electron microscopy. The aim of this chapter is to report the results.

Normal prostate cells can transform into neoplastic cells and form prostate cancer. These neoplastic areas contain cells in various stages of dedifferentiation. Thus some cells are still secreting, whereas others have ceased to produce secretions, although remnants from their earlier activities are sequestered extracellularly [7–9]. Being located outside the cell, prostasomes are an easily accessible target for anti-prostasome antibodies. Considering that localizing metastases of a prostate cancer by, for instance, target-seeking antibodies, would have a strong impact on the therapy of the disease, we also investigated whether the secretory machinery of the early neoplastic cells was similar to that of the

[1]To whom correspondence should be addressed (e-mail ove.nilsson@medcellbio.uu.se).

normal prostate cells and whether prostasome-like structures were still produced and located outside the neoplastic cells.

Materials and methods

The patients comprised 19 men who had been referred to the Department of Urology at Uppsala University Hospital in Uppsala, Sweden, with suspected prostate carcinoma. None of the men had received any hormonal medication before the investigation. They were examined with transrectal ultrasound and core biopsies were taken for routine light and electron microscopy with a ProMag 2.2 biopsy gun (Manan Medical Products, Northbrook, IL, U.S.A.). The biopsies were taken according to a modified sextant protocol including bilateral apex, mid-lateral and base biopsies. These biopsies were sent for routine pathological examination.

For our electron-microscopy study, two additional biopsies were generally taken, each corresponding to a mid-lateral position. These biopsies were fixed with 2.5% glutaraldehyde in PBS, pH 7.2, cut into a few small pieces and embedded in Epon according to conventional techniques.

Based on the pathology report, we selected three patients with benign prostate hyperplasia and four patients with prostate carcinoma with Gleason scores of 7–9 for further studies. From each patient biopsies for two to four plastic blocks were prepared, and each block permitted sectioning at between six and eight different levels of the biopsy for microscopy. Initially, the plastic blocks were cut in 2 μm sections and stained with 1% Toluidine Blue to obtain a view of the cells that were exposed at the sectioned surface of the blocks. Using light microscopy, representative areas with secreting normal cells and secreting neoplastic cells of Gleason grade 1–2 were selected for further studies. The corresponding areas on the plastic blocks were recovered for cutting into 50 nm sections. The sections were placed on slot grids with 0.5% Formvar film, contrasted with lead citrate and uranyl acetate, and examined with an electron miscroscope (Hitachi H-7100). The Independent Ethics Committee of Uppsala University Medical School approved the study.

Results and discussion

The components of the secretory machinery, that is the Golgi apparatus, the storage vesicles and the secretions in the lumen, were structurally similar in both the normal prostate cells and the neoplastic prostate cells. The following description of these components is therefore valid for both cell types.

The Golgi apparatus
Several units of the Golgi apparatus were located around the upper pole of the nucleus. The Golgi membranes showed small budding vesicles and, when these vesicles were released into the cytoplasm, the vesicle membranes became studded with a granulated material (Figure 1). Since the micrographs suggested that the

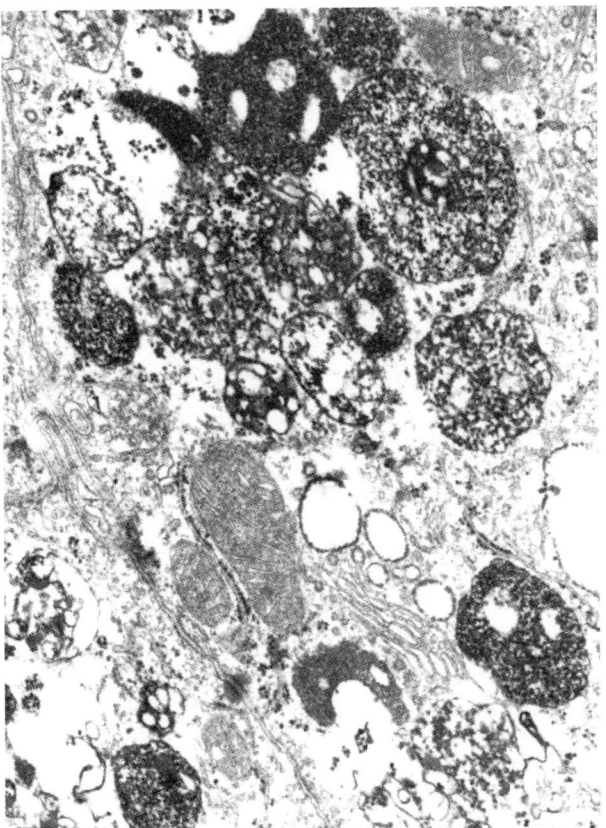

Figure 1

The Golgi area in a cell from a patient with benign prostate hyperplasia

The Golgi membranes are located in the lower right-hand corner. They are releasing Golgi vesicles, which we assume are the early stages of storage vesicles. Initially, the Golgi vesicles are empty, but later they accumulate a dense or granular substance often containing empty rounded areas. Magnification, ×24 700.

vesicles were expanding, we assume that they received materials from the endoplasmic reticulum in the conventional way and therefore grew in size. We regard them as the early stages of storage vesicles. In addition, vesicles with a varying amount of an amorphous substance and a dense, sometimes granular, material were found frequently in the Golgi areas. Our view is that these vesicles were the later developmental stages of the storage vesicles and that they gradually get filled with various components that constitute the contents of mature storage vesicles.

The prostasomes, which are present inside the storage vesicles, ought to be produced at the Golgi apparatus, since we have not observed any transfer of prostasome-like granules from the cytoplasm to storage vesicles. The mode of synthesis of the prostasomes, however, is still not clear. One possible process, as judged from the various structures in the vicinity of the Golgi apparatus, could be the following. The prostasomes are initially synthesized in temporary invaginations of the vesicles (Figure 2). These invaginations contain ribosomes and small parts of the endoplasmic reticulum. Later, the developing prostasomes are seen as

Figure 2

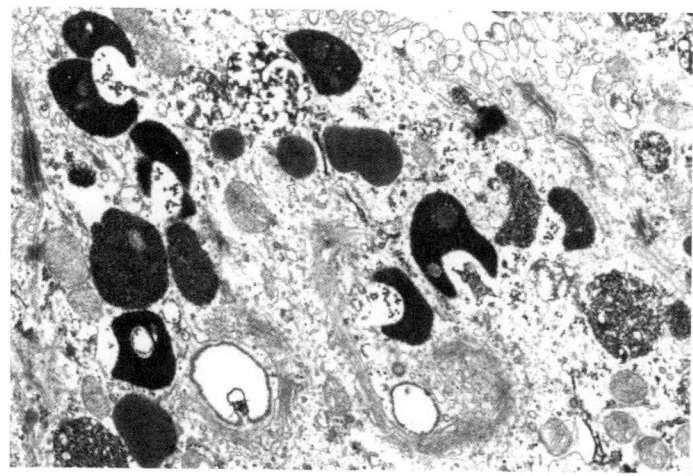

The Golgi area in a cell from a patient with benign prostate hyperplasia

Several of the early developmental stages of the storage vesicles demonstrate invaginations, which often contain small parts of endoplasmic reticulum. Magnification, ×14400.

small light or dark areas inside the early stages of the storage vesicle, where they are continuing their maturation (Figures 1 and 2). Considering the many abilities of the prostasomes, initial synthesis in the cytoplasm seems reasonable. The inclusion of the prostasomes in storage vesicles with their secretory content would be a simple and safe way of securing secretion of prostasomes without forcing them to find their own mechanism of being secreted. If this way of forming prostasomes is true, then our finding of ribosome-like granules in the prostasomes could explain why they contain nucleic acids, probably in the form of double-stranded DNA [10].

The heterogeneous population of vesicles in the Golgi areas also includes other types of cell product, known to be produced in prostate cells. However, the complex ultrastructure of the secretory machinery makes any firm conclusions difficult. With further research we will be able to formulate a more firmly based hypothesis.

Storage vesicles

Storage vesicles and prostasomes were introduced and defined ultrastructurally by Brody et al. [1,2]. The size of the vesicles was calculated to approx. 1 μm and that of the prostasomes to 20–150 nm. It seems that there are two populations of prostasomes; one consists of vesicles with a slightly dense interior and the other is a minor population of larger, denser vesicles with small, dark granules (Figure 3). As yet, the implications of this observation have not been clarified.

The secretions that fill the storage vesicles vary in appearance under the electron microscope, from a dense dark apparition to a light one with only the prostasomes visible (Figure 4). This could be the effect of a maturation process in the vesicles, but it could also depend upon the processing of specimens for

Figure 3

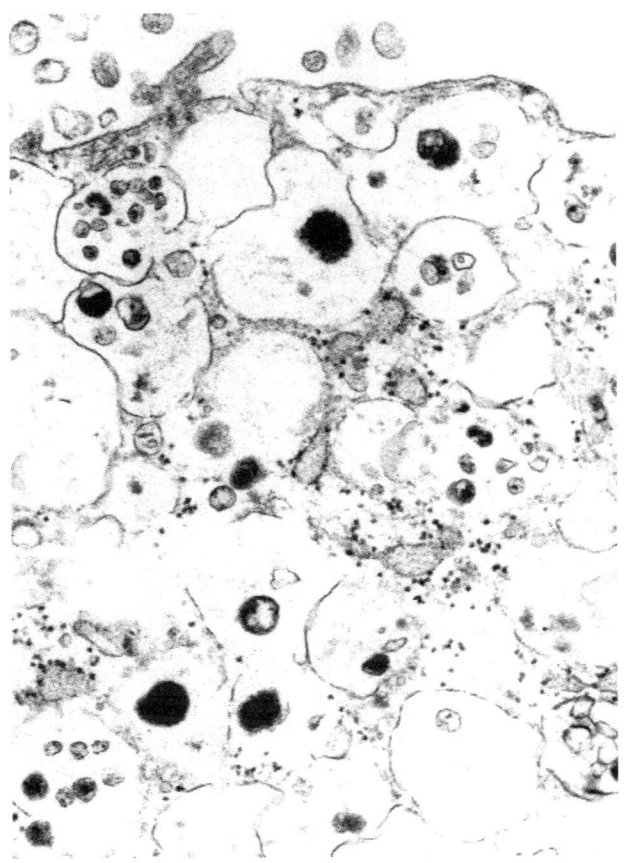

Storage vesicles in the apical part of a cell from a patient with prostate cancer

Since the vesicle membranes are fragile, some have been broken and some have a tortuous course. Prostasomes, but no secretions, are visible in the vesicles. It seems that two types of prostasome are present: small vesicular ones and large dense ones. The latter harbour ribosome-like granules. Magnification, ×31 500.

microscopy, since it is known that the secretions are sensitive to the type of fixation used [7,11].

The discharge of secretions into the lumen is suggested to occur by a budding process, i.e. a release of small blebs from the cells (paracrine secretion), and/or by release of apical vesicles (merocrine secretion). In the latter case, the storage vesicles fuse with the apical cell membrane, thus emptying their contents directly into the lumen [2]. We favour the latter process, since this saves the cells from producing new cell-membrane materials, a task that requires energy. It should be mentioned that tangentially cut sections of bulging apical parts of cells can be misinterpreted as showing the presence of paracrine secretion, and that

Figure 4

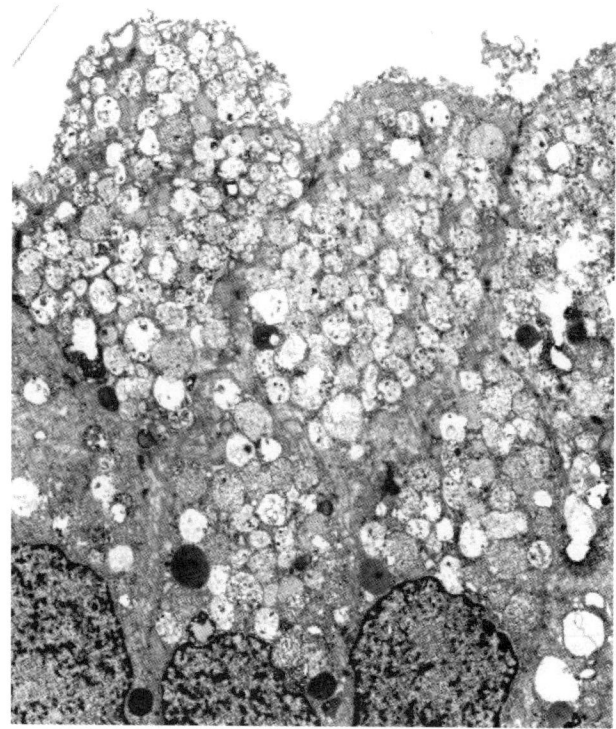

Supranuclear parts of cells from a patient with benign prostate hyperplasia

The secretory cells, which are filled with storage vesicles, have apical surfaces bulging into the duct lumen. Magnification, ×49500.

blebs can be released from cells during apoptosis and after fixation problems, as well as under other conditions.

Secretions in the lumen

The luminal secretions were observed as intermingled areas of a grey amorphous substance and areas with prostasomes (Figure 5). The presence of both vesicle-like and granule-like prostasomes suggests two different types of prostasome, but specific markers for differentiating between the two are required to support this notion.

These results do not reveal any differences in the ultrastructure of the secretory machinery between the normal prostate cell and the well-differentated neoplastic cell; the less well-differentiated cells, however, have lost their ability to produce secretions. Interestingly, patients with prostate cancer can raise anti-prostasome autoantibodies [12]. Normally, when prostasomes are located in the prostate and ejaculatory system, they do not raise any detectable immunological responses. In pathological conditions like metastases, however, when prostasomes are released into other compartments, such as connective tissues and blood vessels, they are evidently able to interact with the immune system. Thus, provided that the neoplastic areas expose spaces with prostasomes to

Figure 5

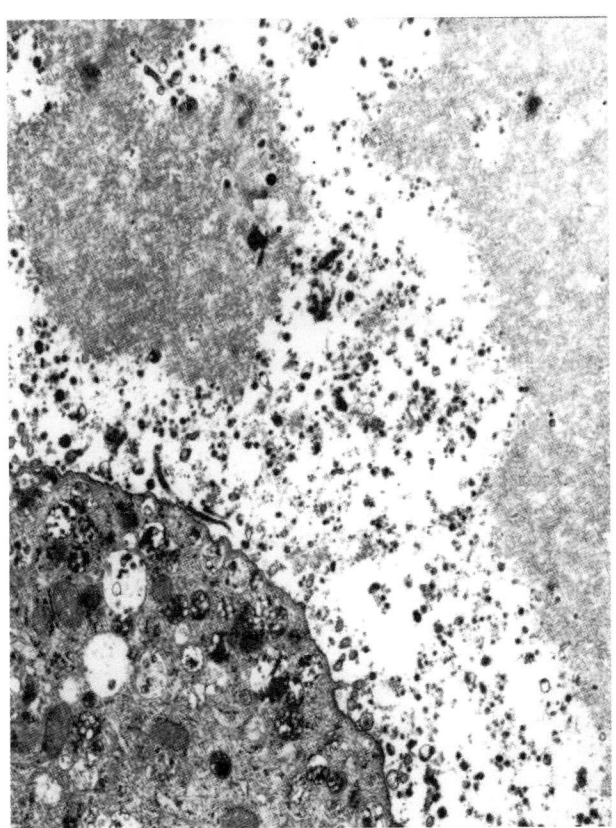

Luminal part of cells from a patient with prostate cancer

The bulging upper part of the cells contains some storage vesicles. In the lumen, many prostasomes and areas with dense secretions can be seen. Magnification, ×26 200.

the immune system, possibilities are offered for both immunodiagnosis and, perhaps, immunotherapy of prostate cancer metastases. Therefore, we are now examining the ultrastructure of metastases from prostate cancers to study this issue.

This study was supported by the foundation Johanna Hagstrand och Sigfrid Linnérs Minne.

References

1   Brody, I., Gottfries, A. and Ronquist, G. (1983) Upsala J. Med. Sci. **88**, 63–80
2   Ronquist, G. and Brody, I. (1985) Biochim. Biophys. Acta **822**, 203–218
3   Stone, M.P., Stone, K.R., Ingram, P., Mickey, D.D. and Paulson, D.F. (1977) Urol. Res. **5**, 185–200
4   Sinha, A.A. and Blackard, C.E. (1973) Urology **2**, 114–120
5   Srigley, J.R. and Hartwick, W.J. (1988) Ultrastruct. Pathol. **12**, 49–65
6   deVries, C.R., McNeal, J.E. and Bensch, K. (1992) Prostate **21**, 209–221
7   Cohen, R.J., McNeal, J.E., Edgar, S.G., Robertson, T. and Dawkins, H.J.S. (1998) Hum. Pathol. **29**, 1488–1494
8   Tannenbaum, M. (1983) Semin. Urol. **1**, 186–192

9   Nilsson, B.O., Egevad, L., Jin, M., Ronquist, G. and Busch, C. (1999) Prostate **39**, 36–40

10   Olsson, I. and Ronquist, G. (1990) Arch. Androl. **24**, 1–10

11   Cohen, R.J., Beales, M.P. and McNeal, J.E. (2000) Hum. Pathol. **31**, 1515–1519

12   Nilsson, B.O., Carlsson, L., Larsson, A. and Ronquist, G. (2001) Upsala J. Med. Sci. **106**, 43–50

3

Prostasomal membrane proteins

Ingrid De Meester[1], Anne-Marie Lambeir and Simon Scharpé
Department of Pharmaceutical Sciences, Laboratory of Medical Biochemistry,
Universiteitsplein 1, B-2610 Antwerp, Belgium

Introduction

The ejaculate has long been considered as a suspension of spermatozoa in a fluid medium, the seminal fluid. It consists of the various secretions of testes and epididymes and of the sex accessory glands (seminal vesicles and the prostate, Cowper's and Littre's glands). About 13–33% of the ejaculate originates from the prostate and this prostatic fluid is slightly acidic, and contains acid phosphatase and remarkably high concentrations of Zn^{2+} [1]. More than 20 years ago, Ronquist and collaborators described the presence of secretory granules and vesicles in human prostatic fluid and seminal plasma. These could be identified after ultracentrifugation of a cell-free seminal plasma and were called prostasomes [2]. Ultrastructurally very similar vesicles were demonstrated within the epithelial cells of the prostate gland [3]. The wide range in organelle size and the nature of the protein/peptide constituents may point to the vesicle population having a heterogeneous nature [4].

Prostasomes are membrane-surrounded organelles with a very unusual membrane-lipid composition. They exhibit a very high cholesterol/phospholipid ratio and contain a large amount of sphingomyelin. Polyunsaturated phosphatidylcholines, common in sperm membranes, are rare in prostasomes [5]. The above-mentioned characteristics, together with a predominance of saturated and mono-unsaturated fatty acids, make the prostasome membrane highly ordered and rigid.

Besides the lipid constituents, a number of proteins have been described in human prostasomes. Quantitatively, prostasomes comprise only 3% of total semen protein [6]. Nevertheless, about 80 different proteins could be distinguished after two-dimensional electrophoresis [4]. This chapter aims to give an overview of the membrane proteins identified in prostasomes. More detailed information on individual membrane proteins is provided elsewhere in this volume.

Since the establishment of hybridoma technology, an ever-increasing number of monoclonal antibodies have been created that has facilitated characterization of cell-surface molecules (originally mainly on leucocytes). Considerable confusion about nomenclature and difficulty with interpretation of results arose as several monoclonal antibodies recognized identical membrane proteins. In a series of international workshops, investigators studied all monoclonal-antibody-defined structures with different techniques and assigned the cell-surface molecules a cluster of differentiation (CD) designation. More than 200 of them have been defined to date. Although a considerable number of the prostasomal membrane proteins have a CD designation, several remain to be clustered.

[1]To whom correspondence should be addressed (e-mail ingrid.demeester@ua.ac.be).

The Protein Reviews On the Web (PROW) service aims to publish concise reports on individual proteins (http://www.ncbi.nlm.nih.gov/PROW/). The latest assignments are retrievable from http://gryphon.jr2.ox.ac.uk. Another useful source of information is the Expert Protein Analysis System (ExPASy; http://www.expasy.ch/) of the Swiss Institute of Bioinformatics. The server is dedicated to the analysis of protein sequences and structures and offers an enzyme subsite with recommendations on enzyme nomenclature.

The Handbook of Proteolytic Enzymes [7], edited by A.J. Barrett, N.D. Rawlings and J.F. Woessner, and the MEROPS database (http://www.merops.co.uk/merops/merops.htm) are valuable resources for comprehensive data in the field. These authors group peptidases on the basis of primary and tertiary structures into families and clans and these are classified further by catalytic mechanism.

Complement regulatory proteins

The complement regulatory proteins CD46, CD55 and CD59 are present on the surface of spermatozoa and they might affect sperm motility [8,9]. In addition, cell-free seminal plasma also contains several complement regulatory proteins. Rooney and colleagues [10,11] demonstrated the glycosylphosphatidylinositol (GPI)-anchored molecule CD59, also known as the membrane-attack-complex-inhibitory protein, in prostasomes. Later, other membrane-bound regulators of the complement system were identified in prostasomes [10–13]. They are listed in Table 1 and include the GPI-anchored CD55 (decay-accelerating factor) and another regulator of the C3 convertase, CD46 (membrane cofactor protein). The ability to remove CD46 from seminal plasma by ultracentrifugation and its presence in the seminal plasma of vasectomized men suggested an origin distal to the ductus deferens [14]. Whereas almost all CD46 was prostasome-bound, a proportion of the CD55 and CD59 remained in the supernatant after ultracentrifugation [11]. Prostasomes are able to transfer their membrane proteins to cells under appropriate conditions and the presence of CD46, CD55 and CD59 may suggest a function in the protection of, for example, spermatozoa against complement attack in the female genital tract.

Miscellaneous CD molecules

In contrast to the molecules described above, CD63, or granulophysin, is a marker for intracellular granules, where it has been used as a marker for late endosomes. It is exported to the cell surface upon activation of platelets, endothelial cells and granulocytes. CD63 is a glycoprotein containing four hydrophobic transmembrane domains with a major extracellular region between transmembrane segments 3 and 4. It belongs to the tetraspanin (or transmembrane 4, TM4) superfamily. Members of this family undergo promiscuous associations with other molecules, including integrins and other tetraspanins. CD63 interacts with the integrins very late antigen (VLA)-3 (CD29/CD49c) and VLA-6 (CD29/CD49f) and the tetraspanins CD9 and CD81. At the cell surface they can

Table 1

CD	Name	Molecular mass (kDa)	Topology	Swissprot accession no.
CD46	Membrane cofactor protein (MCP)	35–40	Type I	P15529
CD52	–	25–29	GPI-anchored	P31358
CD55	Decay-accelerating factor (DAF)	70	GPI-anchored	P08174
CD59	Membrane-attack-complex-inhibitory protein	18–25	GPI-anchored	P13987
CD63	Granulophysin, lysosomal-membrane-associated protein 3	40–60	Four TM domains	P08962

Non-enzymic CD molecules on prostasomes

TM, transmembrane. The data for CD46 are from sperm cells. The molecular mass for CD52 is the apparent molecular mass on SDS/PAGE.

facilitate molecular interactions, increasing the formation and stability of functional signalling complexes [15]. The function of granulophysin on the prostasome membrane is not clear [16].

Human semen contains very potent procoagulant activity. Tissue factor (TF) antigen has been demonstrated in human semen and is found on the prostasome surface, suggesting that the molecule has a prostatic origin [17]. TF is an integral membrane protein and a member of the cytokine receptor superfamily, which functions as a cofactor to enhance activated factor VII proteolytic activity during coagulation. The complex formed by TF and activated factor VII can activate factors X and IX. The crystal structure of the extracellular region of TF has been determined and reveals an extended interface between two barrel-like domains, creating a rather rigid structure [18]. TF is not normally expressed on blood cells or endothelial cells lining the inner vessel wall, but it is present in the vascular adventitia associated with extracellular-matrix proteins. Several epithelia exhibit prominent expression of TF. In general it is expressed on 'barriers' that serve as a so-called 'haemostatic envelope' [19]. Hypothetical functions of seminal TF include protection against abrasion and bleeding during intercourse by rapid clotting at the lesion sites.

Membrane-bound enzymes in prostasomes

Many of the cell-surface CD antigens are ecto-enzymes, catalytic membrane proteins with their active sites outside the cell. They include peptidases, transpeptidases, nucleotidases, phosphodiesterases and phosphatases [20]. A number of them have been identified in prostasomes and in fact these enzymes were among the first prostasomal proteins to be characterized [21–28]. The majority of ecto-enzymes are type-II integral membrane proteins with a short N-terminal cytoplasmic domain, a single transmembrane domain and a large catalytic extracellular domain. However, other membrane topologies are possible (see Table 2). As the nomenclature of ecto-enzymes is very confusing, some frequently used synonyms are given. Each of the two main classes of ecto-enzymes, the nucleotidases and the ectopeptidases, is well represented on prostasomes and will be detailed further below.

In addition, a number of other enzymes have been described on prostasomes. They include phospholipase A_2 [27], lactate dehydrogenase [24], protein kinases [26], alkaline phosphatase [25] and alkaline phosphodiesterase I [25].

Ecto-ATPases/apyrases

Many intracellular processes use ATP as their source of energy and because the enzymes involved consume ATP they are called ATPases. Members of the class of ATPases with their active site outside the cell are known as ecto-(Ca^{2+},Mg^{2+})-apyrases. CD39, an antigen expressed on activated B-cells and natural-killer cells and on a subset of activated T-cells, shares sequence homology with an apyrase and has apyrase activity. This observation led to the identification of a family of ecto-ATPases with related sequences but different membrane topologies and tissue distributions. There is a consensus now that the CD39 family should be referred to as ecto-nucleoside triphosphate diphosphohydrolase (E-NTPDase). Members of the family should be named E-NTPDase1, 2, etc.

CD39 was the first member of this family and it has an unusual topography in that it consists of an intracellular N-terminus, transmembrane domain, a glycosylated extracellular catalytic domain, another transmembrane domain and a C-terminal cytoplasmatic tail. A Ca^{2+}- and Mg^{2+}-dependent ATPase was found on prostasomes [22,28]. This ATPase of the prostasome membrane may be the molecular basis for vectorial transport of Ca^{2+} into prostasomes. To date, no information is available on the molecular nature of the prostasomal ATPases.

5' Nucleotidase

5' Nucleotidase (CD73) catalyses the hydrolysis of phosphate groups from the 5'-carbon of ribose and deoxyribose portions of nucleotides. This widely distributed ecto-enzyme is N-glycosylated and sialylated and is anchored in the membrane via a GPI moiety. The regulation of its expression seems to differ considerably between cell types. 5'-AMP is a preferred substrate and CD73 might serve to regulate the availability of adenosine for interaction with the adenosine receptor [29]. There may be a functional link with adenosine deaminase (ADA), an enzyme that destroys adenosine by converting it to inosine, and with CD26, an ADA-binding protein that is highly expressed on prostasomes [30–32].

Ectopeptidases

Ectopeptidases constitute another group of cell-surface proteins that enable cells to communicate with their environment. Most of the known ectopeptidases have dual existences as membrane-bound and soluble isoforms found in body fluids. The biology of ectopeptidases and their value as disease markers and targets for therapy was reviewed recently [33,34]. Most ectopeptidases are distributed widely. However, their levels of expression are tightly regulated and depend on developmental stage and the activation state of the cells.

The use of specific inhibitors has revealed that some ectopeptidases influence major cell functions, such as metabolic regulation, growth, apoptosis

Table 2

Cluster designation	Names	Molecular mass (kDa)	Topology	Swissprot accession no.
Nucleotidases				
CD39 (EC 3.6.1.5)	Apyrase, ectonucleoside triphosphate-diphosphohydrolase 1, ATP-diphosphohydrolase	70	Two TM domains	P49961
CD73 (EC 3.1.3.5)	Ecto-5'-nucleotidase	69–72	GPI-anchored	P21589
EC 3.1.4.1	Alkaline phosphodiesterase I, plasma-cell membrane-glycoprotein PC-1	120-130	Type II	P22413
EC 3.1.3.1	Alkaline phosphatase, phosphomonoesterase	80–89	GPI-anchored	Q16727
Ectopeptidases				
CD10 (EC 3.4.24.11)	Neprilysin, neutral endopeptidase (NEP)	100	Type II	P08473
CD13 (EC 3.4.11.2)	Aminopeptidase M	150	Type II	P15144
CD26 (EC 3.4.14.5)	Dipeptidyl peptidase IV (DPP IV)	110	Type II	P27487
CD143 (EC 3.4.15.1)	Peptidyl dipeptidase A, angiotensin-converting enzyme (ACE); note: germinal or testicular angiotensin-converting enzyme	170 90	Type I Type I	P12821 P22966

Enzymes on prostasomes

TM, transmembrane.

and motility [34]. The prostasomes are particularly rich in a number of these enzymes [30,35]. Due to its abundance and ease of measurement, aminopeptidase activity [28,36] is even used as a marker enzyme during prostasome purification. Aminopeptidase activity can be transferred from prostasomes to sperm during a fusion process that also transfers lipid [6]. The functional significance of this protein transfer is not yet clear [37]. At least part of the aminopeptidase activity is exerted by CD13, an ectopeptidase that is abundant on a number of epithelial cells and in the haematopoetic system on myeloid cells [37]. Arienti and collaborators [37] found no aminopeptidase activity in the soluble protein fractions from human semen, whereas Huang et al. [38] purified aminopeptidase N from human seminal plasma. However, the supernatant used as a source by these authors (produced at 10 000 g for 20 min) might still have contained prostasome material.

When monoclonal antibodies were raised against prostasomes by intrasplenic immunization, one-third were directed against dipeptidyl peptidase IV (DPP IV) or CD26, a molecule that is abundantly present in human prostasomes [30,32]. Moreover, CD26/DPP IV can be transferred from prostasomes to sperm cells [39]. This membrane-bound serine-type protease removes N-terminal dipeptides from the N-terminus of peptides with proline or alanine at the penultimate position. Using chromogenic or fluorogenic substrates there is a strong preference for a penultimate proline, but this does not always seem to be the case for natural peptide substrates. Some peptides with a penultimate alanine are very good CD26 substrates *in vivo* and even peptides with a penultimate glycine or serine are cleaved under certain conditions [40,41]. A number of possible substrates for CD26 are present in seminal plasma, but the DPP IV activity might be inhibited in this medium by the high concentrations of zinc [1,42,43].

Apart from its enzymic activity, the CD26 antigen has other functions that are perhaps involved in prostasomal physiology. CD26 is known as the ADA-binding protein [31]. ADA also has a second cell-surface receptor, the A_1 adenosine receptor, which is expressed on the sperm surface [44]. In this context, CD26 has been proposed to be part of a 'molecular bridge' (composed of CD26, ADA and the A_1 adenosine receptor) between spermatozoa and prostasomes. On peripheral blood mononuclear cells, HIV gp120 glycoprotein is able to displace endogenous or exogenous ADA from CD26 [45]. Prostasomes might bind and transport HIV particles via membrane-bound CD26.

Plasminogen type 2 (Pg2) binds to the invasive prostate tumour cell line 1-LN. The subtypes Pg2α and Pg2β bind to an L-lysine site-dependent receptor and the highly sialylated Pg2γ, Pg2δ and Pg2ε glycoforms bind primarily to DPP IV. In addition, Gonzalez-Gronow and collaborators [46] have suggested that DPP IV in association with Pg2ε alone regulates expression of pro-matrix metalloproteinase-9 in prostate cancer cells. At present it is not known whether prostasome DPP IV also serves as a Pg2ε receptor and, if so, what the functional implications of this might be.

A polyclonal antiserum raised against purified prostasomes recognized another major antigen of approx. 100 kDa. Using enzymic fragmentation and amino acid analysis, the antigen was identified as neutral endopeptidase (NEP), also known as CD10 or common lymphoid leukaemia antigen (CALLA) [35]. NEP is a widely distributed ectopeptidase identified on normal and malignant cells. It hydrolyses peptide bonds on the amino side of hydrophobic amino acids [47]. Erdös et al. [48] studied NEP in the human male genital tract and found that NEP activities are low in testicular homogenate but high in the particulate epididymes and prostate fractions. NEP from seminal plasma sedimented after ultracentrifugation, which was later confirmed by its prostasomal localization. Glycosylation seems to be important for the correct targeting of NEP to the apical membrane of epithelial cells during intracellular transport in human prostate and for functional activity of the enzyme. Prostatic carcinoma tissue and prostatic carcinoma cells reveal a preferentially cytoplasmic localization of NEP [35]. A number of neuropeptides such as neurotensin and endothelin-1 that are inactivated by NEP are involved in the progression of human prostate carcinoma. NEP on the surface of benign prostate epithelial cells may reduce local concentrations of neuropeptide and regulate their growth-stimulatory capacity [49]. NEP expression is lost in a number of hormone-insensitive, invasive prostate cancer cell lines [50]. This loss of NEP expression results in increased phosphorylation of focal adhesion kinase and also in increased cell migration. Additional experiments revealed that NEP at the cell surface can also induce dephosphorylation of focal adhesion kinase and inhibition of association with other signalling molecules by a mechanism independent of its catalytic activity [49]. Many questions remain concerning the role of NEP in the development of prostate cancer, the clinical correlations between NEP activity/concentration and malignancy and the function of prostasomal NEP.

Seminal fluid also contains another metallopeptidase that uses zinc as a cofactor, namely angiotensin-converting enzyme (ACE, CD143) [51]. Although most of the ACE activity is found in the supernatant after ultracentrifugation, the pelleted material has a very high specific activity [48]. ACE primarily acts as a peptidyl dipeptide hydrolase and is involved in the metabolism of two major vasoactive peptides, angiotensin II and bradykinin. Gene disruption greatly reduces male fertility whereas female mutants were found to be fertile [52]. CD143 exists in a somatic form, which consists of two highly homologous domains that each contain a functional catalytic site [53], and in a germinal form, with one single catalytic site [53,54]. The somatic and germinal forms of CD143 are transcribed from a single gene by two separate promotors. The germinal form is found exclusively in differentiating germinal cells. CD143 on spermatozoa appears to determine the success of the spermatozoa in penetrating the egg. It is not known whether the ACE that is found on prostasomes is transferred to sperm. Both ACE and NEP contribute to the enzymic degradation of bradykinin in sperm [55].

Conclusions

The study of prostasome composition can provide valuable hints concerning the kind of biochemical processes that underly their role, e.g. in safeguarding the reproductive potential of sperm. With the recent expansion of proteomic tools, we can expect even minor protein components to be identified soon. The quantification and measurement of inter-individual variation in key proteins may provide diagnostic and prognostic tools for pathologies of the prostate gland.

In addition, 'whole-organelle' imaging approaches such as immunocyto-chemistry [4,56] and flow cytometry might reveal the presence of undiscovered proteins and may provide a better insight into epitopes that are available for interaction with the environment.

A second step in the elucidation of prostasome function concerns the study of the connectivity between the components and the interactions with molecules that they encounter during their lifespan. Analysis of the dynamic interplay *in vivo* can be considered as a last step in unravelling the involvement of individual proteins in the physiology of these pluripotent organelles [57,58].

References

1   Nieschlag, E. (1998) in Clinical Laboratory Diagnostics (Thomas, L., ed.), pp. 1100–1108, TH-Books Verlagsgesellschaft mbH, Frankfurt
2   Ronquist, G. and Brody, I. (1985) Biochim. Biophys. Acta **822**, 203–218
3   Brody, I., Ronquist, G. and Gottfries, A. (1983) Upsala J. Med. Sci. **88**, 63–80
4   Renneberg, H., Konrad, L., Dammshäuser, I., Seitz, J. and Aumüller, G. (1997) Prostate **30**, 98–106
5   Arienti, G., Carlini, E., Polci, A., Cosmi, E.V. and Palmerini, C.A. (1998) Arch. Biochem. Biophys. **15**, 391–395
6   Arienti, G., Carlini, E., Verdacchi, R. and Palmerini, C.A. (1997) Biochim. Biophys. Acta **1336**, 269–274
7   Barrett, A.J., Rawlings, N.D. and Woessner, J.F. (eds) (1998) The Handbook of Proteolytic Enzymes, Academic Press, London
8   Liszewski, M.K., Farries, T.C., Lublin, D.M., Rooney, I.A. and Atkinson, J.P. (1996) Adv. Immunol. **61**, 201–283
9   Jiang, H. and Pillai, S. (1998) Am. J. Reprod. Immunol. **39**, 243–248
10   Rooney, I.A., Atkinson, J.P., Krul, E.S., Schonfeld, G., Polakoski, K., Saffitz, J.E. and Morgan, B.P. (1993) J. Exp. Med. **177**, 1409–1420
11   Rooney, I.A., Heuser, J.E. and Atkinson, J.P. (1996) J. Clin. Invest. **97**, 1675–1686
12   Kitamura, M., Namiki, M., Matsumiya, K., Tanaka, K., Matsumoto, M., Hara, T., Kiyohara, H., Okabe, M., Okuyama, A. and Seya, T. (1995) Immunology **84**, 626–632
13   Seya, T., Nomura, M., Murakami, Y., Begum, N.A., Matsumoto, M. and Nagasawa, S. (1998) Int. J. Mol. Med. **1**, 809–816
14   Simpson, K.L. and Holmes, C.H. (1994) J. Reprod. Fertil. **102**, 419–424
15   Maecker, H.T., Todd, S.C. and Levy, S. (1997) FASEB J. **11**, 428–442
16   Skibinski, G., Kelly, R.W. and James, K. (1994) Fertil. Steril. **61**, 755–759
17   Fernandez, J.A., Heeb, M.J., Radtke, K.P. and Griffin, J.H. (1997) Biol. Reprod. **56**, 757–763
18   Girard, T.J. (1997) in New Therapeutic Agents in Thrombosis and Thrombolysis (Sasahara, A.A. and Loscalzo, J., eds), pp. 225–260, Marcel Dekker, New York
19   Carson, S.D. and De Jonge, C.J. (1998) J. Androl. **19**, 289–294
20   Goding, J.W. (2000) J. Leukocyte Biol. **67**, 285–311
21   Ronquist, G., Frithz, G. and Jansson, A. (1988) Urol. Int. **43**, 133–138
22   Ronquist, G. (1988) Urol. Int. **43**, 334–340
23   Fabiani, R. and Ronquist, G. (1993) Clin. Chim. Acta **216**, 175–182
24   Olsson, I. and Ronquist, G. (1990) Urol. Int. **45**, 346–349
25   Fabiani, R. and Ronquist, G. (1995) Prostate **27**, 95–101
26   Stegmayr, B., Brody, I. and Ronquist, G. (1982) J. Ultrastruct. Res. **78**, 206–214
27   Lindahl, M., Tagesson, C. and Ronquist, G. (1987) Urol. Int. **42**, 385–389
28   Fabiani, R. (1994) Upsala J. Med. Sci. **99**, 73–112
29   Zimmermann, H. (1992) Biochem. J. **285**, 345–365
30   Vanhoof, G., De Meester, I., van Sande, M., Scharpé, S. and Yaron, A. (1992) Eur. J. Clin. Chem. Clin. Biochem. **30**, 333–338
31   Morrison, M., Vijayasaradhi, S., Engelstein, D., Albino, A. and Houghton, A. (1993) J. Exp. Med. **177**, 1135–1143
32   Schrimpf, S.P., Hellman, U., Carlsson, L., Larsson, A., Ronquist, G. and Nilsson, B.O. (1999) Prostate **38**, 35–39
33   Antczak, C., De Meester, I. and Bauvois, B. (2001) BioEssays **23**, 251–260

34   Antczak, C., De Meester, I. and Bauvois, B. (2001) J. Biol. Regul. Homeostatic Agents **15**, 99–108

35   Renneberg, H., Albrecht, M., Kurek, R., Krause, E., Lottspeich, F., Aumüller, G. and Wilhelm, B. (2001) Prostate **46**, 173–183

36   Stridsberg, M., Fabiani, R., Lukinius, A. and Ronquist, G. (1996) Prostate **29**, 287–295

37   Arienti, G., Carlini, E., Verdacchi, R., Cosmi, E.V. and Palmerini, C.A. (1997) Biochim. Biophys. Acta **1336**, 533–538

38   Huang, K., Takahara, S., Kinouchi, T., Takeyama, M., Ishida, T., Ueyama, H., Nishi, K. and Ohkubo, I. (1997) J. Biochem. (Tokyo) **122**, 779–787

39   Arienti, G., Polci, A., Carlini, E. and Palmerini, C.A. (1997) FEBS Lett. **410**, 343–346

40   De Meester, I., Korom, S., Van Damme, J. and Scharpé, S. (1999) Immunol. Today **20**, 367–375

41   Lambeir, A.M., Proost, P., Durinx, C., Bal, G., Senten, K., Augustyns, K., Scharpé, S., Van Damme, J. and De Meester, I. (2001) J. Biol. Chem., **276**, 29839–29845

42   Ronquist, G. (1998) Int. J. Androl. **21**, 233–234

43   De Meester, I., Vanhoof, G., Hendriks, D., Demuth, H.U., Yaron, A. and Scharpé, S. (1992) Clin. Chim. Acta **210**, 23–34

44   Minelli, A., Allegrucci, C., Mezzasoma, I., Ronquist, G., Lluis, C. and Franco, R. (1999) Biol. Reprod. **61**, 802–808

45   Valenzuela, A., Blanco, J., Callebaut, C., Jacotot, E., Lluis, C., Hovanessian, A.G. and Franco, R. (1997) J. Immunol. **158**, 3721–3729

46   Gonzalez-Gronow, M., Grenett, H., Weber, M., Gawdi, G. and Pizzo, S. (2001) Biochem. J. **355**, 397–407

47   Turner, A.J., Isaac, R.E. and Coates, D. (2001) BioEssays **23**, 261–269

48   Erdös, E.G., Schulz, W.W., Gafford, J.T. and Defendini, R. (1985) Lab. Invest. **52**, 437–447

49   Sumitomo, M., Shen, R., Walburg, M., Dai, J., Geng, Y., Navarro, D., Boileau, G., Papandreou, C., Giancotti, F., Knudsen, B. and Nanus, D. (2000) J. Clin. Invest. **106**, 1399–1407

50   Papandreou, C.N., Usmani, B., Geng, Y., Bogenrieder, T., Freeman, R., Wilk, S., Finstad, C.L., Reuter, V.E., Powell, C.T., Scheinberg, D. et al. (1998) Nat. Med. **4**, 50–57

51   Krassnigg, F., Niederhauser, H., Fink, E., Frick, J. and Schill, W. (1989) Int. J. Androl. **12**, 22–28

52   Krege, J.H., John, S.W., Langenbach, L.L., Hodgin, J.B., Hagaman, J.R., Bachman, E.S., Jennette, J.C., O'Brien, D.A. and Smithies, O. (1995) Nature (London) **375**, 146–148

53   Jaspard, E. and Alhenc-Gelas, F. (1995) Biochem. Biophys. Res. Commun. **211**, 528–534

54   Hubert, C., Houot, A.M., Corvol, P. and Soubrier, F. (1991) J. Biol. Chem. **266**, 15377–15383

55   Heder, G., Böttger, A., Siems, W.E., Rottmann, M. and Kertscher, U. (1994) Andrologia **26**, 295–301

56   Nilsson, B.O., Jin, M., Einarsson, B., Persson, B.E. and Ronquist, G. (1998) Prostate **35**, 178–184

57   Kravets, F.G., Lee, J., Singh, B., Trocchia, A., Pentyala, S.N. and Khan, S.A. (2000) Prostate **43**, 169–174

58   Kelly, R.W. (1999) Int. J. Androl. **22**, 2–12

4

Prostasomal membrane granulophysin

Grzegorz Skibinski

Department of Clinical Biochemistry, Queen's University, Grosvenor Road, Belfast BT12 6BJ, Northern Ireland, U.K.

Introduction

There are three major means by which organisms transfer substances outwards: (i) small molecules move by diffusion directly through the plasma membrane or via membrane channels or carriers (lipids, water, inorganic ions and small organic molecules); (ii) large molecules or small molecular aggregates, encased in membrane granules or vesicles, are extruded from the cell by exocytosis, a process in which the vesicle membrane fuses with the apical cell membrane (as a result, the membrane-bound content is then discharged into the environment); and (iii) structures that are secreted by the cell in membrane-bound form (including mucous granules in molluscs, hepatic phospholipid vesicles of mammals and various products of aposecretion and holosecretion). Prostasomes belong to the last category.

The presence of cytoplasmic granules or vesicles is a feature shared by the cells of many tissues. These structures contain biologically active substances that, after release into the extracellular milieu, play an important role in cell and tissue function (reviewed in [1]). They can also act as messengers and perform their functions at a distance from the cell from which they originated. One approach that has been taken to elucidate the function of vesicles is to study the biochemical composition of vesicles and of vesicle membranes in particular. Pioneering studies were performed in cells of neural or endocrine origin. Several integral membrane proteins of secretory vesicles have been defined, such as synaptophysin [2,3], synaptobrevin [4,5] and chromogranins/secretogranins [6,7]. An interesting integral membrane protein named granulophysin has been identified in the granules of human platelets [8,9]. It was found that granulophysin is deficient in platelets from patients with the Hermansky–Pudlak syndrome, a condition in which abnormal platelet function is associated with hypopigmentation and the storage of ceroid in reticuloendothelial cells [9]. Immunoelectron microscopic examination in platelets confirmed that granulophysin is indeed a granule membrane protein [10]. Anti-granulophysin monoclonal antibodies were obtained and used to explore the distribution of granulophysin epitopes in different tissues. ELISA, immunohistochemistry and Western blotting were used in such studies. The results revealed a wide distribution of granulophysin epitopes in human tissues and cells. Particularly notable was the presence of the granulophysin epitope in all of the endocrine glands and in most of the exocrine glands and their ducts, including parotid gland, exocrine pancreas and prostate. In addition, the anti-granulophysin antibodies recognized proteins in skeletal-muscle cells and in epithelial lining cells. The granular pattern of staining observed in most cells suggested that in these tissues, as in platelets, the anti-granulophysin monoclonal antibody-reactive epitope was also associated with intracellular granules. In contrast to other membrane integral proteins of secretory vesicles, granulophysin was not limited to the cells of neural or endocrine origin and was found in most tissues tested [11]. It was also reported that anti-granulophysin monoclonal antibodies recognized molecules in the membranes of granules or vesicles of endothelial cells, neutrophils, lymphokine activated killer cells and the U937 monocytic cell line [8,12].

Granulophysin is expressed in human prostasomes and prostate

The observations mentioned above prompted my co-workers and I to study whether monoclonal antibodies against granulophysin would react with prostasomes purified from human semen. At the time when these studies were undertaken a marker that would enable us to identify prostasomes or even purify them from semen was needed urgently.

Although originally designed to study particles the size of cells and nuclei, the techniques of flow cytometry have been applied successfully to analysis and sorting of much smaller particles. The rich variety of subcellular and microscopic particles studied by flow cytometry now includes chromosomes, micro-organisms, chloroplasts, plankton and endosomes [13]. Our initial experiments established that prostasomes stained with fluorochrome-conjugated antibodies could also be studied by flow cytometry [14]. Granulophysin epitope was reproducibly found on 60–85% of particles in all prostasome preparations studied. Negative staining was obtained using antibodies against CD46 (membrane cofactor protein), CD62 (platelet α-granule marker) and CD16 (granulocyte marker; see Table 1). The anti-granulophysin antibody did not react with capacitated or non-capacitated sperm cells. A negative reaction with anti-CD46 antibody was surprising because the presence of membrane cofactor protein in particulate matter of semen has already been described [15]. Also, lack of staining with sperm cells came as a surprise because prostasomes have been shown to associate with sperm cells previously [16]. However, we cannot exclude the possibility that prostasomes associate weakly with sperm and become dissociated during processing.

In Western blots, anti-granulophysin antibody defined molecules of a slightly lower molecular mass than in platelet-dense granules (Figure 1). When considering antigens in seminal plasma it is necessary to be aware of the possible influence that prostatic enzymes may have, as major alterations of some seminal-plasma antigens have been observed. It should also be noted that anti-granulo-physin antibody reacted strongly with epithelial lining cells in prostate and epididymal sections [17]. Data from electron-microscopic studies published by Ronquist and Brody suggested that prostasomes are present within storage

	Fluorescence (%)				**Table 1**
Antibody	Prostasome preparation...	1	2	3	HL 60 cells
Anti-granulophysin (D545)		79.35 (117)	61.64 (171)	85.34 (126)	98.10 (148)
Anti-CD46		9.28 (61)	9.01 (52)	6.42 (61)	
Anti-CD62		2.61 (62)	1.34 (51)	3.43 (48)	
Anti-CD16		1.32 (41)	1.05 (40)	1.15 (35)	

Granulophysin expression in prostasomes: flow-cytometric studies

Prostasomes from three different preparations were used. Prostasomes and cells were incubated with an optimal dilution of antibody for 60 min at 4°C, washed twice and reacted with sheep anti-mouse IgG F(ab')$_2$ fragments conjugated with FITC for 30 min at 4°C. Finally, prostasomes were washed and fixed in 1.5% formaldehyde and analysed by flow cytometry (EPICS flow cytometer; Coulter, Hialea, FL, U.S.A.). Results are presented as percentages of fluorescent particles, with mean fluorescence values given in parentheses. HL60 cells were permeabilized by treatment with cold ethanol before staining. The anti-granulophysin mouse monoclonal antibody, D545, was a kind gift from Dr Gerrard of Manitoba Institute of Cell Biology, Winnipeg, Canada. Other antibodies were obtained from commercial sources.

vesicles in cells of the prostatic acinar epithelium [18]. We have found that granulophysin molecule is expressed in both the prostate and the epididymis. It is therefore possible that prostasomes derive from more than one organ. In fact, vesicles present in seminal plasma of other species have been shown to originate from seminal vesicles and epididymis (see other chapters in this volume). It has been observed recently that anti-granulophysin antibody recognizes the same or similar proteins to CD63. Anti-CD63 antibodies effectively blocked detection of the protein by anti-granulophysin using immunofluorescence, ELISA, immunoblotting and FACS analysis. N-terminal sequencing over the first 37 amino acids revealed that granulophysin shares identity with CD63, melanoma antigen ME491 and pltg40 [19].

Concluding remarks

A legacy of molecular evolution is the formation of gene families that encode proteins often serving related functions. One such family gaining recent attention is the tetraspanin superfamily, whose membership has grown to nearly 20 known genes since its discovery in 1990 [20]. CD63 is a member of this family. All genes in the family encode cell-surface proteins that span the membrane four times, forming two extracellular loops. They have been conserved during evolution and are present in mice, humans and *Drosophila* [21,22]. The biological function of tetraspanins is unknown. Awaiting definitive functional studies, we can only put together pieces of a puzzle that has been built by raising antibodies against these proteins and looking at their distribution, associations and functions. Data accumulated so far indicate that some tetraspanins are found in virtually all tissues (CD81, CD82, CD9 and CD63), whereas others are highly restricted, such as CD37 (B-lymphocytes) or CD53 (lymphoid and myeloid cells). Many of these proteins have a tendency for promiscuous associations with other molecules,

Figure 1

A   B

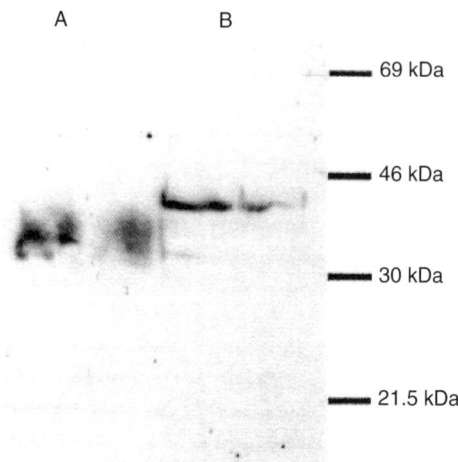

— 69 kDa

— 46 kDa

— 30 kDa

— 21.5 kDa

Western-blot analysis of solubilized whole-cell homogenates of human platelets (lane B) and prostasomes (lane A)

Samples were electrophoresed under reducing conditions using a 12% gel according to the method of Laemmli [26]. After transfer to Immobilon membrane, the membrane was stained with the D545 (anti-granulophysin) antibody and developed using the enhanced chemiluminescence system. Positions of the molecular-mass markers are indicated on the right. Reprinted with permission from the American Society for Reproductive Medicine (Fertility and Sterility 1994, 61, 755–759).

including lineage-specific proteins, integrins [21,23] and other tetraspanins [24]. In terms of function, they are involved in diverse processes such as cell activation and proliferation, adhesion and motility, differentiation and cancer. It is possible that these functions may all relate to their ability to act as 'molecular facilitators', grouping specific cell-surface proteins, and thus increasing the formation and stability of functional signalling complexes.

Future studies

Whether granulophysin/CD63 on prostasomes serves any specific extracellular functions remains to be determined. Several possibilities seem worth addressing. First, it may play a role in the transmission of signals to neutrophils, as shown by Jy et al. [25], for interactions between neutrophils and platelets. Secondly, it may protect the membrane by being resistant to various proteolytic enzymes on the basis of the presence of large complex carbohydrates, including poly-n-acetyl-lactosamines. Finally, it may present carbohydrate in a functional way as a ligand for adhesion molecules. The presence of other tetraspanin family members on prostasomes should also be determined.

References
1   Deyrup-Olsen, I. and Luchtel, D.L. (1998) Int. Rev. Cytol. **183**, 95–141
2   Lu, B. and Chow, A. (1999) J. Neurosci. Res. **58**, 76–87

3   Stridsberg, M. (1995) Upsala J. Med. Sci. **100**, 169–199

4   Mochida, S. (2000) Neurosci. Res. **36**, 175–182

5   Bentz, J. and Mittal, A. (2000) Cell Biol. Int. **24**, 819–838

6   Aunis, D. and Metz-Boutigue, M.H. (2000) Adv. Exp. Med. Biol. **482**, 21–38

7   Borges, R., Machado, J.D., Alonso, C., Brioso, M.A. and Gomez, J.F. (2000) Adv. Exp. Med. Biol. **482**, 69–81

8   Gerrard, J.M., Lint, D., Sims, P.J., Wiedmer, T., Fugate, R.D., McMillan, E., Robertson, C. and Israels, S.J. (1991) Blood **77**, 101–112

9   Shalev, A., Machaud, G., Israels, S.J., McNicol, A., Singhroy, S., McMillan, E., Greenberg, A.H., White, J.G., Witkop, C.J. and Gerard, J.M. (1992) Blood **80**, 1231–1237

10   Israels, S.J., Gerrard, J.M., Jacques, Y.V., McNicol, A., Cham, B., Nishibori, M. and Bainton, D.F. (1992) Blood **80**, 143–152

11   Hatskelzon, L., Dalal, B.T., Shalev, A., Robertson, C. and Gerrard, J.M. (1993) Lab. Invest. **68**, 509–519

12   Abdelhaleem, M., Hatskelzon, L., Dalal, B., Gerrard, J.M. and Greenberg, A.M. (1991) J. Immunol. **147**, 3053–3059

13   Hammond, T.G., Majewski, R.R., Morre, D.J., Schell, K. and Morrissey, L.W. (1993) Cytometry **14**, 411–420

14   Skibinski, G., Kelly, R.W., Harkiss, D. and James, K. (1992) Am. J. Reprod. Immunol. **28**, 97–103

15   Thaler C.J., Critser, J.K., McIntyre, J.A. and Faulk, W.P. (1989) Fertil. Steril. **52**, 463–468

16   Ronquist, G., Nilsson, B.O. and Hjerten, S. (1990) Arch. Androl. **24**, 147–157

17   Skibinski, G., Kelly, R.W. and James, K. (1994) Fertil. Steril. **61**, 755–759

18   Ronquist, G. and Brody, I. (1985) Biochim. Biophys. Acta **822**, 203–218

19   Nishibori, M., Cham, B., McNicol, A., Shalev, A., Jain, N. and Gerrard, J.M. (1993) J. Clin. Invest. **91**, 1775–1782

20   Horejsi, V. and Vlcek, C. (1991) FEBS Lett. **288**, 1–4

21   Schmidt, C., Kunemund, V., Wintergerst, E.S., Schmitz, B. and Schachner, M. (1996) J. Neurosci. Res. **43**, 12–31

22   Kopczynski, C.C., Davis, G.W. and Goodman, C.S. (1996) Science **271**, 1867–1870

23   Berditchevski, F., Bazzoni, G. and Henler, M.E. (1995) J. Biol. Chem. **270**, 17784–17790

24   Radford, K.J., Thorne, R.F. and Hersey, P. (1996) Biochem. Biophys. Res. Commun. **222**, 13–18

25   Jy, W., Mao, W.W., Horstman, L., Tao, J. and Ahn, Y.S. (1995) Blood Cells Mol. Dis. **21**, 217–221

26.   Laemmli, U.K. (1970) Nature (London) **227**, 680–685

5

Anti-prostasome antibodies: a new marker for prostate cancer

Anders Larsson*[1], Lena Carlsson*, B. Ove Nilsson† and Gunnar Ronquist*
*Department of Medical Sciences, Clinical Chemistry, University Hospital, SE-751 85 Uppsala, Sweden, and †Department of Medical Cell Biology, Unit of Anatomy, Biomedical Center, P.O. Box 571, SE-751 23 Uppsala, Sweden

Introduction

Prostate-specific antigen (PSA) is considered to be the golden standard for the detection of prostate cancer and has both high sensitivity and specificity for prostate cancer [1]. However, for many patients an elevated PSA value does not have an impact on mortality and morbidity. The life expectancy of 70–80 year-old males with elevated PSA values and highly differentiated prostate cancer was not significantly different from a matched group of males without prostate cancer, even after follow-up periods of 10 years or more [2]. It is thus important to find new markers that can differentiate between benign and metastasing forms of prostate cancer. To be able to metastase, the tumour has to be in direct contact with a blood vessel. The PSA molecule is smaller than a prostasome and is therefore more likely to diffuse through the interstitial space. In contrast, prostasomes are so large that diffusion in the interstitial space is probably limited. To be able to detect prostasomes in the blood, we believe that the tumour has to be in direct contact with blood vessels. Potentially, this could result in a stronger correlation between metastasis and prostasomes than between metastasis and PSA.

Similar to tumour cells, the concentration of prostasomes in the blood is low and thus difficult to detect. We have tried to detect prostasomes in plasma by flow cytometry, which is well suited for clinical use. We used FITC-labelled chicken anti-prostasome antibodies to distinguish the prostasomes from the background. Unfortunately, we were not able to detect prostasomes with this method. An alternative to flow cytometry is detection of mRNA in the blood by PCR. However, at present, PCR is not sufficiently automated for rapid testing of a large number of samples. We have instead tried to develop an assay for the detection of anti-prostasome antibodies in patient serum.

Knowledge regarding the potential use of assays for detection of prostasomes or antibodies to prostasomes can be drawn from previous studies on p53. The *p53* suppressor gene is implicated in the control of the cell cycle, DNA synthesis and repair as well as apoptosis [3,4]. Mutations in this gene are one of the most common genetic alterations in human malignancies and are found in approx. 50% of non-small-cell lung cancer tumours and in 90% of small-cell lung carcinomas [5–7]. In most tissues the protein product of wild-type *p53* is expressed only for very short periods. Mutated p53, on the other hand, often has a considerably longer half-life than the wild-type protein [8] and is therefore more readily detected in tumour cells by immunohistopathological methods. The concentration of p53 in serum is too low to be able to be detected by ELISA or radioimmunoassay. Instead, antibodies against the p53 protein can be detected in sera from patients with cancer. A question of great importance is whether the presence of antibodies is associated with p53 mutations, and a connection between the presence of antibodies against the p53 protein in sera and p53 mutations has been reported [9–11].

During the last few years there have been reports on antibodies to p53 in serum from patients with ovarian cancer [12], breast cancer [13], head and neck cancer [14] and various other carcinomas [15]. Increased levels of anti-p53 antibodies were associated with reduced survival in patients with breast cancer [13] and in head and neck cancer [14], but had no impact on survival in ovarian carcinoma [12]. The relationship between the presence of p53 antibodies and prognosis seems to vary between different studies and different diagnoses. In one study we have shown that autoantibodies to p53 are associated with poor prognoses in patients with colorectal cancers [16]. We have also shown that anti-p53 antibodies are associated with increased life expectancy in non-small-cell lung cancer patients treated with radiotherapy [17].

Why was the expression of antibodies in patients who were going to receive radiotherapy a good prognostic sign in our study? One explanation for improved survival of patients expressing p53 antibodies might be that these patients have a more efficient immune system that is more capable of counteracting the tumour disease. The presence of antibodies might in itself have a favourable impact on survival, perhaps by a direct effect on tumour cells expressing the antigen. If so, improved survival might be expected in patients participating in a study currently in progress in the U.S.A., in which patients are vaccinated with p53 peptides in a phase-III clinical trial sponsored by the National Cancer Institute (see http://clinicalstudies.info.nih.gov/detail/A_1999-C-0137.html).

If tumour patients develop antibodies to p53 it is also plausible that patients with prostate cancers will develop antibodies to prostasomes. A prostasome is an antigen, which would usually evoke a strong immune response. Previous studies on infertile patients have shown that these patients may have antibodies to prostasomes [18]. We have also produced chicken antibodies and mouse monoclonal antibodies against prostasomes, which is further evidence that they are immunogenic. We chose to immunize chickens with prostasomes, as chicken antibodies are more closely related to human autoantibodies than mammalian antibodies [19]. Chicken antibodies also have other advantages over mammalian antibodies: they do not activate the human complement system and do not interact with human Fc receptors, rheumatoid factors or anti-mouse IgG antibodies (HAMA). The antibodies were purified from egg yolk, thus circumventing the bleeding procedure necessary to produce mammalian polyclonal antibodies. A single egg yolk contains more than 100 mg of IgY and a hen lays approx. 20 eggs/month. A hen thus produces considerably more antibodies than a rabbit. Chicken antibodies react with several of the prostasomes' antigens and they also react with different types of prostasome.

We also made monoclonal antibodies, which were tested against microtitre plates coated with prostate tissue, metastasis prostasomes and seminal prostasomes. Bound antibodies were detected with an enzyme-labelled anti-mouse IgG conjugate. The monoclonal antibodies showed greater differences than the chicken antibodies when tested against the different prostasomes. This indicates that prostasomes differ in protein composition depending on their source. Three of these antibodies were characterized by Western blotting. Monoclonal antibody 8H10 detected a protein band with a molecular mass of approx. 100 kDa, whereas antibody 8C7 detected a 60 kDa band and 3D7 detected a 30 kDa band. The exact identity of these bands is presently unknown, but the 60 kDa band corresponds to the molecular mass of the complement-regulatory membrane cofactor protein (CD46). We also tested patient sera against SDS/PAGE-separated prostasomes and blotted them on to nitrocellulose membranes [20]. The patterns differed between patients but 60 and 8 kDa bands were seen in several patients. The variation in pattern between patients indicates that it is probably better to use a mixture of prostasome proteins to coat the microtitre plates. In the initial experiments we thus used prostatic-tissue prostasomes.

Anti-prostasome ELISA [20]

An ELISA was used to detect anti-prostasome antibodies in serum. Plates (F96, Polysorp, Nunc) were coated with 4 μg of purified prostasomes, isolated from human prostatic tissue, diluted in 100 mM $NaHCO_3$, pH 9.5 (coating buffer), for 2 h at 37°C. The plates were then washed and blocked for 1 h at 37°C with the coating buffer containing 3% BSA. After blocking, the plates were washed three times with 200 μl of PBS containing 0.1% Tween 20 (PBS-T) and incubated with 200 μl of patient or control serum, diluted 1:50 in PBS, for 2 h at 37°C. After three further washes with 200 μl of PBS-T, 100 μl of goat anti-human IgG horseradish peroxidase-conjugated antibodies, diluted 1:1000 in PBS, were added and incubated for 1 h at room temperature. The plates were washed three times with 200 μl of PBS-T and incubated with substrate (tetramethyl benzidine; Zymed Laboratories, San Francisco, CA, U.S.A.) for 15 min at room temperature while being protected from light. The reaction was stopped by adding 50 μl of 1.8 M H_2SO_4. Absorbance was measured at 450 nm in an ELISA reader (SPECTRAMax 250; Molecular Devices, Sunnyvale, CA, U.S.A.).

Results

All patients tested had detectable levels of autoantibodies to prostasomes, whereas all controls showed background values, and there was no overlap between the groups. There was no significant difference between males and females in the control group. The range of absorbance values (measured at 450 nm) for the control group was 0.03–0.15 (0.097±0.04; mean±S.D.), whereas all patients had absorbance values that were higher than the controls, i.e. 0.23–0.34 (0.278±0.03; Figure 1).

Figure 1

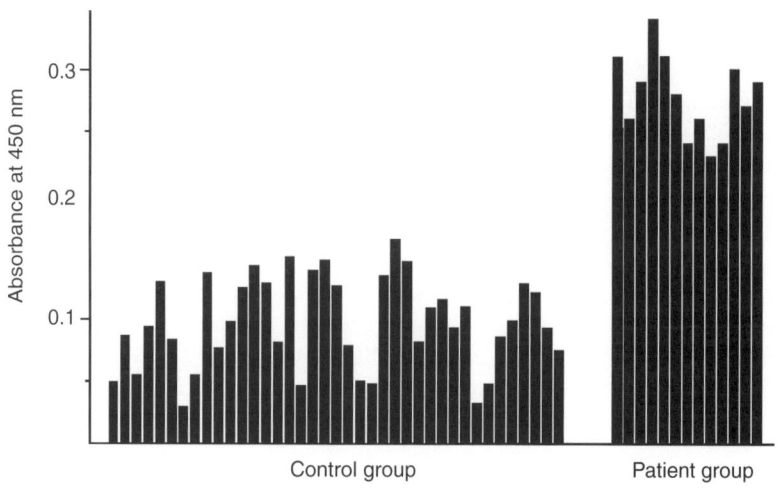

Anti-prostasome ELISA absorbance values of 39 controls and 13 men with prostate cancer

Reproduced from [20] with permission from Taylor and Francis Ltd.

We also tested the antibody response to metastasis prostasomes in the same assay. The pattern was similar, but the assay had higher background levels. This was probably due to autoantibodies bound to the prostasomes that were not removed during purification of the prostasomes. By changing the capture antigen in the ELISA it may be possible to increase the specificity of the assay and possibly differentiate between different types of prostate cancer.

Conclusion

Patients with prostate cancer have antibodies to prostasomes. Further studies have to be carried out to optimize the capture antigen and to evaluate the prognostic value of the assay.

References

1 Barry, M.J. (2001) N. Engl. J. Med. **344**, 1373–1377
2 Adolfsson, J. (2000) Läkartidningen **97**, 3870–3874
3 Suzuki, H., Takahashi, T., Kuroishi, T., Suyama, M., Ariyoshi, Y., Takahashi, T. and Ueda, R. (1992) Cancer Res. **52**, 734–736
4 Chiba, I., Takahashi, T., Nau, M.M., D'Amico, D., Curiel, D.T., Mitsudomi, T., Buchhagen, D.L., Carbone, D., Piantadosi, S., Koga, H. et al. (1990) Oncogene **5**, 1603–1610
5 Levine, A.J., Momand, J.J. and Finlay, C.A. (1991) Nature (London) **351**, 453–456
6 Mitsudomi, T., Hamajima, N., Ogawa, M. and Takahashi, T. (2000) Clin. Cancer Res. **6**, 4055–4063
7 Takahashi, T., Nau, M.M., Chiba, I., Birrer, M.J., Rosenberg, R.K., Vinocour, M., Levitt, M., Pass, H., Gazdar, A.F. and Minna, J.D. (1989) Science **246**, 491–494
8 Elledge, R.M. (1996) J. Natl. Cancer Inst. **88**, 141–143
9 Winter, S.F., Minna, J.D., Johnson, B.E., Takahashi, T., Gazdar, A.F. and Carbone, D.P. (1992) Cancer Res. **52**, 4168–4174

10   Davidoff, A.M., Iglehart, J.D. and Marks, J.R. (1992) Proc. Natl. Acad. Sci. U.S.A. **89**, 3439–3442

11   Crawford, L.V., Pim, D.C. and Bulbrook, R.D. (1982) Int. J. Cancer **30**, 403–408

12   Angelopoulou, K., Rosen, B., Stratitis, M., Yu, H., Solomou, M. and Diamandis, E.P. (1996) Cancer **78**, 2146–2152

13   Peyrat, J.P., Bonneterre, J., Lubin, R., Vanlemmens, L., Fournier, J. and Soussi, T. (1995) Lancet **345**, 621–622

14   Bourhis, J., Lubin, R., Roche, B., Koscielny, S., Bosq, J., Dubois, I., Talbot, M., Marandas, P., Schwaab, G., Wibault, P., Luboinski, B. et al. (1996) J. Natl. Cancer Inst. **88**, 1228–1233

15   Angelpoulou, K., Diamandis, E.P., Sutherland, D.J., Kellen, J.A. and Bunting, P.S. (1994) Int. J. Cancer **58**, 480–487

16   Kressner, U., Larsson, A., Bergström, R., Påhlman, L., Glimelius, B. and Lindmark, G. (1998) Br. J. Cancer **77**, 1848–1851

17   Bergqvist, M., Brattström, D., Larsson, A., Holmertz, J., Hesselius, P., Rosenkvist, L., Wagenius, G. and Brodin, O. (1998) Anticancer Res. **18**, 1992–2002

18   Carlander, D., Stålberg, J. and Larsson, A. (1999) Upsala J. Med. Sci. **104**, 179–190

19   Allegrucci, C., Ronquist, G., Ove Nilsson, B., Carlsson, L., Lundqvist, M., Minelli, A. and Larsson A. (2001) Am. J. Reprod. Immunol. **46**, 211–219

20   Nilsson, B.O., Carlsson, L., Larsson, A. and Ronquist, G. (2001) Upsala J. Med. Sci. **106**, 43–50

6

Prostatic secretory mechanisms and the formation of prostasomes

**Gerhard Aumüller*[1], Jürgen Seitz*, Jian Song*, Martin Albrecht*,
Anders Bjartell† and Beate Wilhelm***
*Department of Anatomy and Cell Biology, University of Marburg, Robert-Koch-Str. 6,
D-35033 Marburg, Germany, and †Department of Urology, Malmö University Hospital,
Malmö, Sweden

Introduction

The concept of prostasomes, as developed by Ronquist and his associates in the 1980s, has allowed for a better understanding of sperm–semen interactions. The mechanism of prostasome formation, however, is far from being clear. In the present chapter this issue is discussed with respect to the mechanism of apocrine secretion and release, operative in the rat coagulating gland and human prostate, respectively. The contribution of neutral endopeptidase (NEP) present in the apical cell membrane of human prostatic secretory cells during the apocrine release of secretory material is demonstrated by immunohistochemistry. By immunoelectron microscopy, we show immunoreactivities for chromogranin A (CgA; derived from prostatic neuroendocrine cells), and for semenogelin (derived mainly from seminal vesicles) present in prostasomes. These are therefore interpreted as complex secondary products from different genital glandular sources forming in the seminal pathways. The designation as 'seminosomes', therefore, seems more appropriate for these particles. The mechanism of apocrine secretion is elucidated, with the rat coagulating gland serving as a model. Apocrine-secreted proteins synthesized in the rat coagulating gland share the following peculiarities. Their biosynthesis and post-translational modification (including an unusual form of glycosylation) take place in the cytoplasm, and intracellular transport proceeds without participation of the endomembrane system, the Golgi apparatus and secretion granules. It is hypothesized that blood-serum-derived trans-sudated albumin entering the secretory cells may function as a carrier of the apocrine-secreted proteins. Only parts of these features apply to secretion in the human prostate, namely apocrine release. The release of prostatic secretory proteins occurs via both a merocrine and an apocrine extrusion mode, often via apical protrusions, surrounded by the plasma membrane containing NEP. Mixing of these protrusions with the secretions from other parts of the genital tract is thought to result in the extracellular formation of the seminosomes.

[1]To whom correspondence should be addressed (e-mail aumuelle@mailer.uni-marburg.de).

History

The cell biology of secretion has been studied preferentially in salivary glands and the pancreas [1]. Results obtained in these merocrine secreting glands have been generalized for most other glandular organs [2]. This has been questioned during the last few years, especially with regard to the glands related to the reproductive organs [3].

The fruitful concept of prostasomes, as developed by Ronquist, Brody and collaborators, was based on their ultrastructural findings in secretory particles derived from human prostatic fluid of healthy men as well as infertile patients [4–6]. They described the 'prostasomes' as granular particles, ranging in size from 20 to 150 nm, surrounded by tri-, penta- or multi-lamellar membranes. Prostasomes isolated by preparative ultracentrifugation and Sephadex G-200 chromatography manifested characteristic enzyme activities associated with their surrounding membranes [7]. The secretion mechanism of prostasomes has been studied by Brody et al. [5], who termed the secretory vacuoles present in human prostatic secretory cells as 'storage vesicles', containing the granular prostasomes.

Two different modes of release of secretory material from human prostatic glandular cells have been distinguished by the authors of these studies. (i) In the first, prostasomes are delivered into the lumen by merocrine exocytosis, starting with the fusion of adjacent membranes belonging to the storage vesicle and the apical plasma membrane and the discharge of the contents into the lumen, which is followed by a recycling of the vacuolar membrane. (ii) In the second, the whole storage vesicle is translocated from the interior of the cell into the acinar lumen through the plasma membrane, a secretion mode named 'diacytosis' by Brody et al. [5].

Non-classical secretion

Apart from the classical secretory proteins, for example those typical of the pancreas [1,8], Muesch et al. [9] have listed a number of proteins, some of which have calcium- and phospholipid-binding capacities, that are exported from the cytosol without being proteolytically processed and several of which lack hydrophobic sequences or are N-terminally acylated. As has been pointed out by Seitz et al. [10] and Steinhoff et al. [11], these characteristics apply perfectly to the secretory transglutaminase secreted in an apocrine fashion in the rat coagulating gland. Similar observations have been made at other locations of the male genital tract.

Manin et al. [12] have studied the apocrine secretion of the mouse vas deferens protein, a major androgen-dependent protein of mouse deferential fluid. They showed by immunoelectron microscopy that mouse vas deferens protein immunoreactivity was distributed in the cytoplasm; it was never detected in the cisternae of the endoplasmic reticulum, Golgi apparatus or secretion granules, but was abundant in apical protrusions and intraluminal fluid. A computer-assisted analysis of the protein structure showed the lack of a signal peptide for translocation by the endoplasmic reticulum. As the protein was also found on the sperm surface [13], the authors inferred a specific role of this apocrine-secreted protein in the surface-mediated regulation of sperm functions. This interpretation agrees well with recent findings of other apocrine-secreted proteins in the coagulating gland, such as carboanhydrase II [14] and, in the human genital tract, perhaps 5′-nucleotidase [15]. This mode of release shares some features with apocrine secretion, described in sex-related glands as early as 1922 by the German histologist Paul Schiefferdecker ([16]; see [3] for historical details).

Functional organization of prostatic epithelium

The functional differentiation of the prostate starts during puberty, at the age of 11–14 years, and is characterized by the general development of pseudostratified epithelium with basal and secretory cells in all parts of the gland, an increase in the diameter of the lumen of the ducts, and the formation of secondary and tertiary ramifications of the glandular ducts, pushing the interstitial stroma aside. CgA-immunoreactive neuroendocrine cells, presumably derived from the neural crest and invading urethral epithelium via the surrounding mesenchyme, concentrate in the periurethral portion, from where they are distributed throughout the more peripheral portions of the glandular epithelium [17–19]. The formation of mature glands starts at about 15–16 years, being easily recognized from the immunoreactivity of the glandular adluminal cells for prostate-specific antigen (PSA) and other secretory proteins. The epithelium of the mature gland consists of typical basal cells and columnar adluminal cells, displaying prostatic acid phosphatase (PAP), PSA or β-microseminoprotein immunoreactivities, as well as a few neuroendocrine cells containing 5-hydroxytryptamine (serotonin) and/or calcitonin.

The apical border of the secretory cells usually bulges into the lumen (Figure 1a) and is replenished by several large vacuoles, some containing reticular to granular material or even being electron-translucent [17]. The diameter of the vacuoles ranges between 600 and 1200 nm with a mean of 1000 nm. The granules are sometimes surrounded by a single membrane, which may be absent or replaced by reticular material. They measure between 60 and 320 nm, averaging approx. 100 nm (Figure 1b). Only a few cytoplasmic organelles, mostly lysosomes, multivesicular bodies, mitochondria and cytoskeletal fibres, are interspersed between the vacuoles. The perinuclear portion of each cell contains the Golgi apparatus, a number of mitochondria, short sections of rough endoplasmic reticulum and cytofilaments, as well as electron-dense bodies, resembling lysosomes [17].

Co-localization studies

We have used mono- and polyclonal antibodies directed against PAP, PSA, α_1-anti-chymotrypsin (ACT), protein C inhibitor (PCI), human albumin, semenogelin, actin, tissue-type transglutaminase and NEP. These were applied to both semi-thin and ultra-thin sections of epon-embedded human prostatic samples as well as to isolated prostasomes (provided kindly by Professor G. Ronquist, Department of Medical Sciences, University Hospital, Uppsala, Sweden), fixed at low glutaraldehyde concentrations. Double-immunolabelled preparations were evaluated by conventional light microscopy, confocal laser

Figure 1

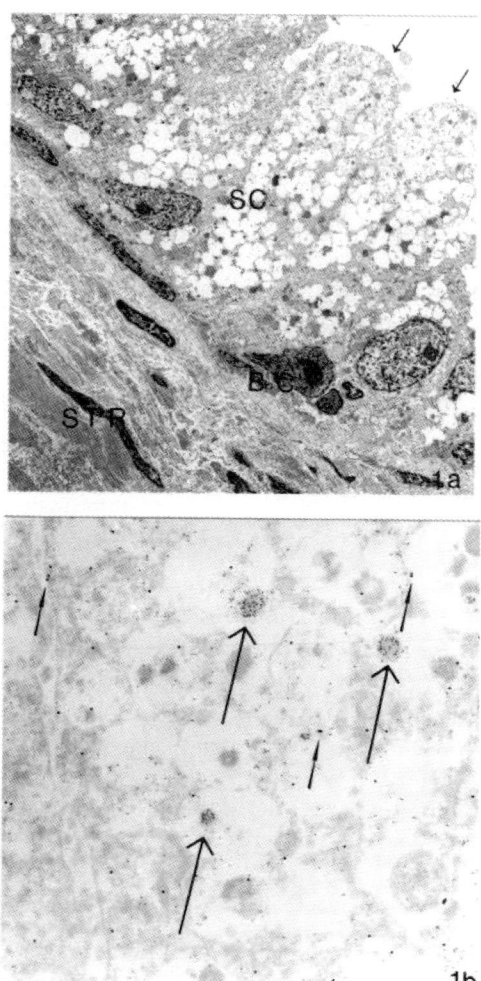

(a) Survey electron micrograph of human prostatic secretory cells and (b) high-power electron micrograph of human prostatic secretory granules and vacuoles

(a) STR, stroma; BC, basal cell; SC, secretory cell; arrows, apical domes. (b) Human prostatic secretory granules and vacuoles were immunogold-labelled for PSA (10 nm particles; large arrows) and α_1-anti-chymotrypsin (20 nm particles; small arrows). Magnification: (a) ×4000; (b) ×20000.

scanning microscopy (courtesy of Professor E. Weihe, University of Marburg, Marburg, Germany) and transmission electron microscopy.

Whereas PAP and PSA were co-localized in about 95–97% of the normal prostatic secretory cells (Figure 2a), only partial co-localization could be obtained for both ACT and PCI. These proteins were restricted to a relatively small proportion of prostatic glandular cells (<40%). ACT immunoreactivity was nevertheless present in the secretory compartment of the cells (apical protrusions), whereas PCI was found

Figure 2

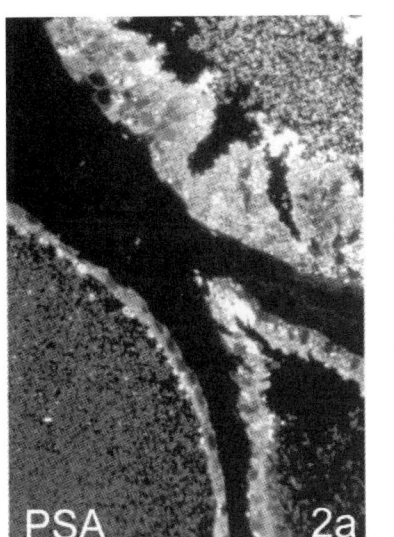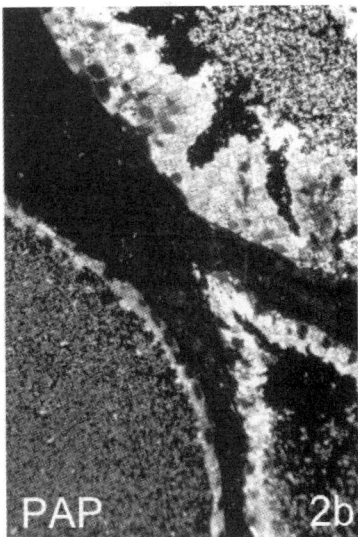

Confocal laser scanning micrograph of (a) PSA and (b) PAP co-localization in human prostatic epithelium

Magnification: ×360.

preferentially in the intermediate portion of prostatic epithelium, perhaps the Golgi apparatus. No semenogelin immunoreactivity (with the exception of ejaculatory duct cells) was found in prostatic epithelium (although recently a low semenogelin signal has been detected in prostatic secretory cells; Å. Lundwall, personal communication), and CgA immunoreactivity was confined to the neuroendocrine cells. Tissue-type transglutaminase and albumin immunoreactivities were found in very few secretory cells and showed only partial co-localization with the secretory proteins. NEP immunofluorescence was present in the apical plasma membrane of the secretory cells, where it was associated closely to actin immunofluorescence (Figure 2b) No NEP immunoreactivity was found in basal and neuroendocrine cells. Ultrastructurally, NEP immunoreactivity was restricted to the outer leaflet of the apical plasma membrane, preferentially close to the base of short microvilli and the apical domes, bulging into the lumen. Likewise, it surrounded the protrusions found inside the lumen. NEP therefore is an excellent marker for the release of apical protrusions containing secretory proteins in the human prostate.

Quantitative evaluation of the labelling density of the secretory and cytoplasmic compartments of prostatic glandular cells showed slightly divergent distribution patterns for PSA, PAP, ACT and PCI. Differences were partly dependent on the fixation type used (formaldehyde and glutaraldehyde concentrations, additives such as acrolein, picric acid and Ca^{2+}). A high labelling density was obtained with the monoclonal anti-PSA antibody, which bound preferentially to the secretory granules and the cytoplasm. The vacuoles and the residual cell body were labelled only weakly. A comparable distribution pattern was obtained with the polyclonal PAP antibody, which showed a complex binding pattern over the membranes of the vacuoles and their particular contents. Vacuoles and granules were only slightly labelled with the polyclonal ACT antibody, which presented a stronger cytoplasmic labelling. The latter pattern was also typical for the polyclonal PCI antibody.

Malatesta et al. [20] have recently published a study on extraprostatic PSA synthesis in the human placenta, in which they found PSA immunoreactivity over the cytoplasm and cytoplasmic protrusions and the microvilli, which are thought to represent the sites of release of PSA from the syncytiotrophoblast cell.

Our labelling patterns point to largely simultaneous processing of PAP and PSA, whereas ACT and PCI immunoreactivities are more pronounced in particular cells, perhaps exhausted or newly differentiated forms. Their heterogeneous distribution (in pathologically altered specimens) is discordant with the relatively homogeneous pattern of PSA and PAP immunoreactivities. In normal prostatic epithelium a granular ACT immunoreactivity was present in the apical protrusions.

Immunoelectron microscopy on Lowicryl-embedded prostasomes, provided kindly by Professor G. Ronquist, as well as conventional glutaraldehyde- and osmium-fixed prostasome specimens from our laboratory showed differential labelling patterns with the various antibodies used. The most important finding was that a rather high labelling density in prostasomes was achieved when a polyclonal antibody against semenogelin was applied (courtesy of Professor H. Lilja, Malmö University Hospital, Malmö, Sweden); semenogelin is the characteristic secretion product of seminal vesicles. About 90% of the particles present were labelled, and background labelling was negligible. Less-dense labelling of the particles was achieved with antibodies against ACT and PAP, whereas only low labelling densities were obtained with antibodies directed against PSA, PCI, NEP and CgA (Table 1). Obviously, the labelling density is at least partly dependent on the antibody used and does not necessarily reflect the true distribution ratios of the different proteins. More important, however, is the fact that the prostasomes do not contain exclusively prostatic secretory material, but also a proportion derived from other sources in the genital tract, mainly the seminal vesicles and perhaps also the epididymis.

Terminology: aposomes, prostasomes and seminosomes

Apocrine secretion is both an innervation-dependent and hormonally regulated release mechanism [21], occurring in its typical form in the dorsal prostate and coagulating gland of the rat (for review, see [3]). Proteins synthesized in a non-classical fashion, lacking a signal peptide and being glycosylated outside the Golgi apparatus, flow to the apical compartment of the cells [10]. Increasing amounts of blood-serum-borne albumin (thought to act as a carrier) and newly synthesized plasma-membrane proteins are required to form apical blebs that are filled with the secretory proteins [15] and which are released into the lumen with the active participation of cytoskeletal proteins, at least actin, gelsolin and myosin. Released blebs are called aposomes and represent the primary corpuscular secretory elements. In the coagulating gland, the apocrine and merocrine release mechanisms occur simultaneously but are obviously regulated separately.

Table 1

Antibody	Low-grade glutaraldehyde-fixed, Lowicryl-embedded specimen	High-grade-glutaraldehyde/osmium-fixed, Epon-embedded specimen
PSA	+	+/−
PAP	++	+
ACT	+	+/−
PCI	+/−	+/−
CgA	+	+
SG I	++	+/−
NEP	+	+

Immunoelectron microscopic labelling densities of prostasomes

++, Strong labelling; +, weak labelling; +/−, background labelling; SG I, secretogranin I.

A slightly divergent release mechanism is observed in human epididymal epithelium, seminal vesicles and prostate gland. In the epididymis, the apical plasma membrane contains numerous stereocilia, which may serve as modified aposomes, i.e. they are released into the lumen, where they form vesicular structures. Glycosylphosphatidylinositol-anchored membrane proteins present in these vesicles bind to the surface of spermatozoa [22,23]. The same is true for the apical cell domes or blebs, which are frequently, but not necessarily generally, seen in seminal vesicles and the prostate epithelium. These domes, containing several vacuoles with or without dense granules, bulge into the lumen. In tangential sections they appear as isolated bleb-like particles. As yet, the participation of the cytoskeleton in the release mechanism of these domes is not clear. We presume that these domes are the equivalents of aposomes, disintegrating inside the lumen, there releasing their vacuolar and granular contents. In addition to this apocrine type of secretion, typical merocrine release of granules and of vacuoles is observed, in human prostate gland and seminal vesicles as well as in the mouse vas deferens [13]. There is no doubt that the prostate contributes a considerable proportion of the proteinaceous content of the prostasomes.

The question arises of whether or not prostasomes are the equivalents of prostatic secretion. Their size and electron-microscopic appearance reminds one of the granules present in prostatic secretory vacuoles and were the reasons for naming them as prostasomes. Nevertheless, there are two aspects that argue against such a contention. (i) Frequently, prostasomes are surrounded by tri- to penta-lamellar membranes or even have a multilamellar appearance, whereas prostatic granules are usually surrounded by a single membrane and are often free of any surrounding structure. (ii) The only granules present in prostatic cells resembling the membrane-bound appearance of the prostasomes are the endocrine, presumably 5-hydroxytryptamine-containing, granules of the prostatic neuroendocrine cells. As CgA and other components of the neuroendocrine granules have been described in prostasome fractions, their prostatic origin, despite them being from combined exocrine and endocrine secretions, would infer that these particles were prostate-specific. It would mean, however, that these particles are secondary formations, due to a mixing of (neuro)peptides released from the endocrine cells and purely exocrine secretory components, such as PSA and PAP.

As we have shown by immunohistochemistry and immunoelectron microscopy, prostasomes also contain proteins secreted by the seminal vesicles, such as semenogelin. It is therefore more appropriate to name them seminosomes, as they are apparently composite secondary products formed after mixing of secretions from different sources of the human male genital tract. This does not interfere with their eminent significance in a variety of divergent functions, such as sperm-motility stimulation, anti-bacterial activity and sperm–membrane interactions.

Conclusions

Our immunohistochemical and immunoelectron microscopic studies on prostatic glandular cells as well as isolated prostasomes (and comparisons with seminal vesicles, Cowpers gland and epididymis, which are not described here) provide circumstantial evidence that prostasomes are secretion-containing particles that form secondarily after the release of the individual components from the respective glands, in both apocrine and merocrine fashions. The term seminosomes would therefore be more appropriate for these particles.

We gratefully acknowledge the excellent technical help of I. Dammshäuser, G. Hoffbauer and M. Dreher. Antibodies were donated kindly by Professor Hans Lilja and Professor Johan Malm (semenogelins) of Malmö University Hospital; prostasome specimens embedded in Lowicryl were provided kindly by Professor Gunnar Ronquist, Department of Medical Sciences, University Hospital, Uppsala, Sweden.

References

1   Palade, G. (1975) Science **189**, 347–358
2   Darnell, J., Lodish, H. and Baltimore, D. (1986) in Molecular Cell Biology, pp. 105–130, Scientific American Books, New York
3   Aumüller, G., Wilhelm, B. and Seitz, J. (1999) Ann. Anat. **181**, 437–446
4   Brody, I., Ronquist, G., Gottfries, A. and Stegmayr, B. (1981) Scand. J. Urol. Nephrol. **15**, 85–91
5   Brody, I., Ronquist, G. and Gottfries, A. (1983) Upsala J. Med. Sci. **88**, 63–80
6   Ronquist, G. and Brody, I. (1985) Biochim. Biophys. Acta **822**, 203–218
7   Ronquist, G., Brody, I., Gottfries A. and Stegmayr, B. (1978) Andrologia **10**, 261–272
8   Rothman, J.E. and Orci, L. (1992) Nature (London) **355**, 409–415
9   Muesch, A., Hartmann, E., Rohde, K., Rubartelli, A., Sitia, R. and Rapoport T.A. (1990) Trends Biochem. Sci. **15**, 86–88
10   Seitz, J., Keppler, C., Rausch, U. and Aumüller, G. (1990) Histochemistry **93**, 525–530
11   Steinhoff, M., Eicheler, W., Holterhus, P.M., Rausch, U., Seitz, J. and Aumüller, G. (1994) Eur. J. Cell Biol. **65**, 49–59
12   Manin, M., Lecher, P., Martinez, A., Tournadre, S. and Jean, C. (1995) Biol. Reprod. **52**, 50–62
13   Taragnat, C., Berger, M. and Jean, C. (1990) J. Androl. **11**, 279–286
14   Wilhelm, B., Meinhardt, A., Renneberg, H., Linder, D., Gabius, H.-J., Aumüller, G. and Seitz, J. (1999) Eur. J. Cell Biol. **78**, 256–264
15   Aumüller, G., Renneberg, H., Schiemann, P.-J., Wilhelm, B., Seitz, J., Konrad, L. and Wennemuth, G. (1997) in The Fate of the Male Germ Cell, Advances in Bioscience, vol. 39 (Ivell, R. and Holstein, A.-F., eds), pp. 193–219, Plenum Press, New York
16   Schiefferdecker, P. (1922) Die Hautdrüsen des Menschen und der Säugetiere, ihre biologische und rassenanatomische Bedeutung, sowie die Muscularis sexualis, E. Schweizerbart, Stuttgart
17   Aumüller, G. (1979) in Handbuch der mikroskopischen Anatomie des Menschen, vol. VII, part 6 (Oksche, A. and Vollrath, L., eds), pp. 1–380, Springer, Berlin
18   Aumüller, G., Leonhardt, M., Janssen, M., Konrad, L., Bjartell, A. and Abrahamsson, P.-A. (1999) Urology **53**, 1041–1048
19   Aumüller, G., Leonhardt, M., Renneberg, H., von Rahden, B., Bjartell, A. and Abrahamsson, P.-A. (2001) Prostate **46**, 108–115

20 Malatesta, M., Mannello, F., Luchetti, F., Marcheggiani, F., Condemi, L., Papa, S. and Gazzanelli, G. (2000) J. Clin. Endocrinol. Metab. **85**, 317–321

21 Satoh, Y., Habara, Y., Kanno, T. and Ono, K. (1993) Cell Tissue Res. **274**, 1–14

22 Kirchhoff, C. and Hale, G. (1996) Mol. Hum. Reprod. **2**, 177–184

23 Vernet, P., Faure, J., Dufaure, J.-P. and Drevet, J.R. (1997) Mol. Reprod. Dev. **47**, 87–98

7

Prostasome-like vesicles and sperm interaction in stallions

Alba Minelli*[1], Lavinia Liguori*, Cinzia Allegrucci†, Cesare Castellini‡, Rafael Franco§ and Gunnar Ronquist¶

*Dipartimento di Scienze Biochimiche e Biotecnologie Molecolari, Sezione di Biochimica Cellulare, Via del Giochetto, 06123 Perugia, Italy, †Division of Animal Physiology, School of Biosciences, University of Nottingham, Sutton Bonington Campus, Loughborough LE12 5RD, U.K., ‡Dipartimento di Scienze Zootecniche, Università di Perugia, Perugia, Italy, §Department Bioquímica i Biologia Molecular, Facultat de Química, Martí i Franquès 1, 08028 Barcelona, Spain, and ¶Department of Medical Sciences, Clinical Chemistry, University Hospital, SE-751 85 Uppsala, Sweden

Introduction

Membrane vesicles have been identified in the seminal plasma of several mammals [1–5]. These vesicles are secreted from different accessory organs of the repoductive system and are named after the producing organ. Prostasomes have been identified in human ejaculate [3], vesiculosomes in bovine seminal plasma [4] and membrane organelles of epididymal origin in rabbit, ram and rat ejaculates [1,2,5]. These extracellular vesicles express different proteins and enzymes on their surfaces and are involved in several physiological roles. They have immunosuppressive activity [6,7], anti-bacterial and growth-inhibitory effects [8,9], and an enhancing effect on sperm-cell motility [10]. These vesicles also exert opposing effects in the fertilization process in different species: in rabbits fertilization is inhibited [11], whereas in cattle and humans forward sperm motility is promoted and the acrosome reaction induced [12,13].

This chapter will summarize data dealing with the occurrence of vesicles in horse seminal plasma and the occurrence of a fusion-like process between vesicles and sperm cells. This process can modify horse cells' enzymic activities, thereby providing the vesicles with new physiological roles in the fertilization process.

Materials and methods

Horse semen was obtained using an artificial vagina from stallions of proven fertility stabled at ARAM (Associazione Regionale Allevatori Marche), Macerata, Italy. Horse seminal vesicles were obtained according to Ronquist and Brody [3]. Horse vesicles and sperm cells were also fixed according to Ronquist and Brody [3] and ultrastructural examinations performed by Centro di Microscopia Elettronica, Università di Perugia, Perugia, Italy. Immunofluorescence analyses

Figure 1

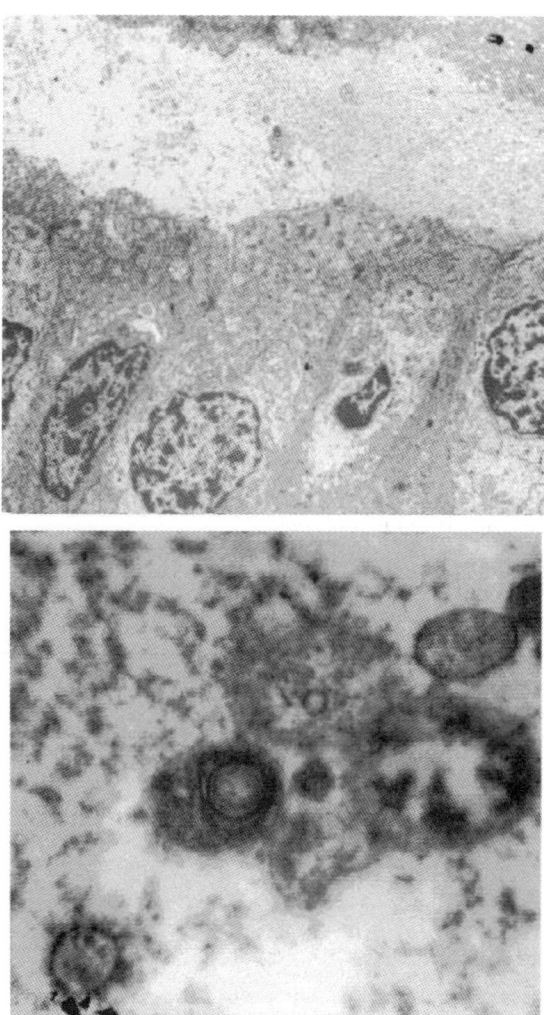

Ultrastructural images of horse prostate gland

were performed with a Leica TCS 4D confocal scanning laser microscope attached to an inverted Leitz microscope.

PC21 antibody (anti-A_1 adenosine receptor) was provided by the laboratory of R. Franco. Endopeptidase activity was assayed according to Laurell et al. [14]. Diadenosine polyphosphate hydrolase activity was assayed by determining the residual amount of substrate by HPLC [15].

Figure 2

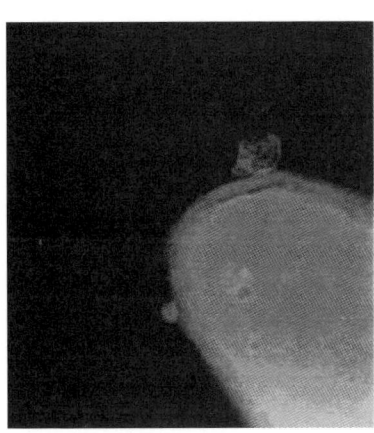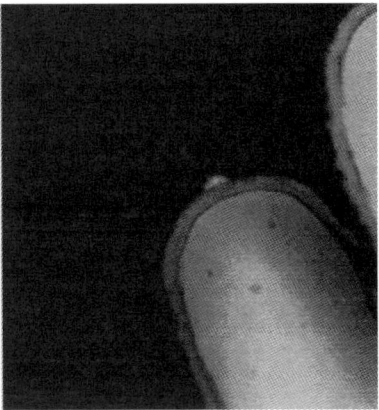

Scanning electron micrograph of horse sperm cells and prostasome-like vesicles during the fusion-like process

Results and discussion

Horse seminal plasma was used to obtain membrane vesicles by ultracentrifugation at 105000 g for 120 min. The pellet was then chromatographed on a Sephadex G-200 column and fractions were examined for absorbance at 280 nm and endopeptidase activity as the marker enzyme for prostasomes [16].

Electron-microscopic analyses showed that the pellet consisted mainly of vesicles that were roughly round in shape, with a diameter ranging between 75 and 150 nm. The particles were surrounded by a bilaminar unit membrane and contained amorphous matrix and electron-dense material. These particles showed membrane-bound 5'-nucleotidase (7.3 nmol/min per mg of protein), endopeptidase (3.4 nmol/min per mg of protein), dipeptidyl peptidase IV (6.8 nmol/min per mg of protein), alkaline phosphatase (8.4 μmol/h per mg of protein) and diadenosine polyphosphate hydrolase (85 nmol/min per mg of protein) activities.

The determination of nucleosides and nucleotides in the horse vesicles showed that only adenosine and its tri- and di-phosphate derivatives were present, whereas AMP and other non-adenylic nucleotides were absent. The addition of these vesicles to horse sperm cells affected the adenylic metabolism so that the intracellular ATP concentration was maintained at a high level and AMP was not accumulated. Consequently the energy charge of the cells was stabilized at vital physiological values, in contrast to the situation observed with sperm cells alone. After 60 min of incubation in 0.188 M Hepes buffer containing 0.3 M glucose, 8.33 mM lactose and 50 units of penicillin, pH 7.4, the energy charge values, ranging between 0.5 and 0.4, were indicative of the deficient vitality of the cells.

The vesicles identified in horse seminal plasma presented nucleotidic components [17] and enzyme activities similar to those found in human prostasomes [18,19]. Moreover, secretory vesicles were found in horse prostate gland (Figure 1). Therefore, there is a strong possibility that these horse particles

Figure 3

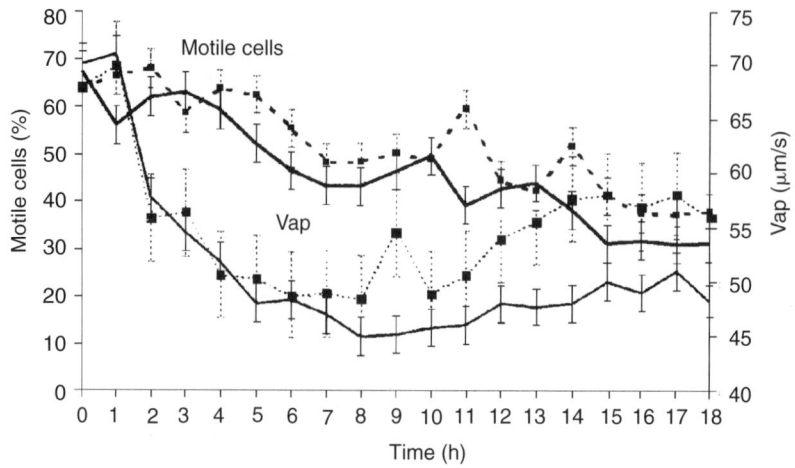

Kinetic parameters of horse sperm cells as analysed by computer-assisted semen analyser

The percentage of motile cells (%, left-hand axis) and the mean velocity of the smooth cell path (μm/s, right-hand axis) are shown. The solid lines indicate the absence of vesicles; the dotted and dashed lines show the addition of vesicles.

Figure 4

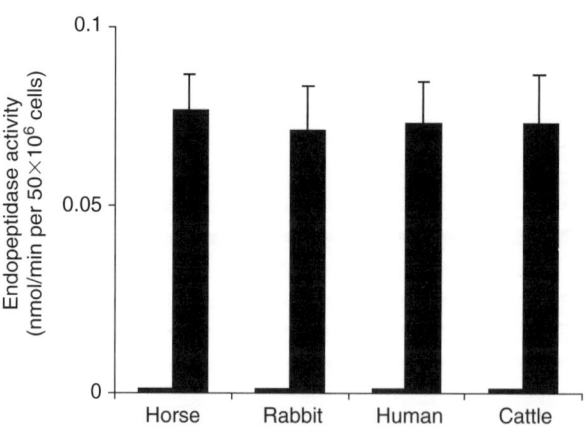

Transfer of endopeptidase activity from horse prostasome-like vesicles to mammalian sperm cells

Left-hand bars for each species represent endopeptidase activities prior to a 60 min incubation with prostasome-like vesicles; right-hand bars show activities after incubation.

may be of prostatic origin, although further investigations will be needed to confirm their anatomical origins.

It has been hypothesized that prostasomes can assist the fertilization potential of spermatozoa by adhering to them, thus modifying their microenvironment [13]. The interaction between horse sperm cells and vesicles was investigated by electron microscopy and biochemical studies. The ultrastructural data showed that the interaction is a fusion-like process (Figure 2): the event started with the formation of a bridge between the two membranes, i.e. those of the sperm cells and vesicles, and then proceeded gradually until the vesicle was embedded in the sperm-cell membrane. The fusion-like process occurred at pH 7.5, which is in contrast to the fusion at strongly acidic pH reported between prostasomes and human spermatozoa [20].

The addition of the vesicles had few effects on the kinetic parameters of horse sperm cells, as analysed by computer-assisted semen analyser (Figure 3). The pecentage of motile cells did not increase significantly, whereas the average velocity of the smooth cell path was increased significantly after 6 h of incubation with the vesicles. Because of its late onset, this effect was considered to be of little physiological relevance.

Endopeptidase activity was found on the membranes of washed horse spermatozoa that had been incubated previously with prostasome-like vesicles. This enzyme activity, which is a marker of horse seminal vesicles, is not normally present on horse sperm cells. These cells acquired the enzyme by undergoing the fusion-like process with the vesicles that were added. The transfer was time-dependent: horse spermatozoa acquired endopeptidase activity of 0.07 nmol/min per 50×10^6 cells after a 60 min of incubation with prostasome-like vesicles at 37°C. Prostasome-like vesicles transferred their endopeptidase activity to sperm cells of other species, which means that the fusion-like process is not species-specific (Figure 4). Rabbit, human and bull spermatozoa showed endopeptidase activity after a 60 min incubation in the presence of the vesicles.

Figure 5

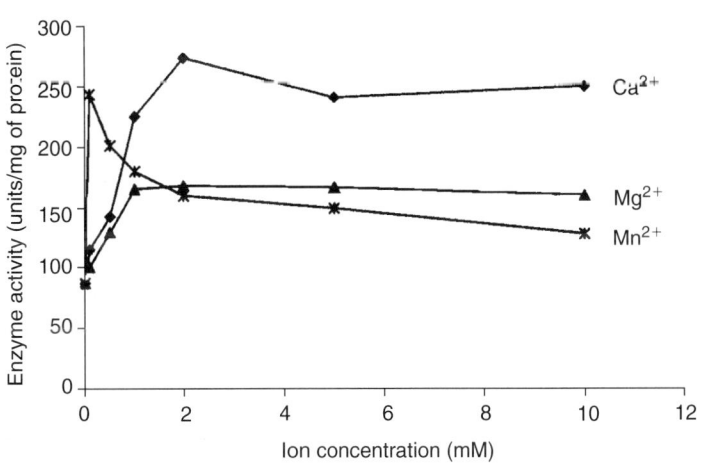

Effect of bivalent cations on diadenosine polyphosphate hydrolase activity

Table 1	Time (min)	Diadenosine tetraphosphate (μM)	ATP (μM)	ADP (μM)	AMP (μM)	Adenosine (μM)
	0	500	0	0	0	0
	20	420	44	170	46	215
	40	300	45	250	56	278
	60	290	45	270	46	403
	90	220	41	305	41	496
	120	200	27	300	31	565

Time-course degradation of diadenosine tetraphosphate by the diadenosine polyphosphate hydrolase activity of prostasome-like vesicles

Prostasome-like vesicles showed the ability to degrade diadenosine polyphosphate compounds. These compounds have several physiological effects, ranging from the regulation of cell proliferation, differentiation and apoptosis to signalling through specific receptors [21–27]. Prostasome-like ecto-diadenosine polyphosphate hydrolase is a glycosylphosphatidylinositol-anchored protein, as shown by treatment of vesicles with phosphatidylinositol-specific phospholipase C, which resulted in the loss of the hydrolytic activity of the vesicles.

The enzymic activity of diadenosine polyphosphate hydrolase was affected by bivalent cations (Figure 5). The enzyme's hydrolytic activity was increased by Ca^{2+}, Mg^{2+} and Mn^{2+}. Ca^{2+} and Mg^{2+} caused maximal stimulation at 1–2 mM, whereas Mn^{2+} had its maximal stimulating effect at 0.1 mM. Therefore, Mn^{2+} is an activating ion that is 10–20-fold more effective than Ca^{2+} or Mg^{2+}. The activity of diadenosine polyphosphate hydrolase was pH-dependent (Figure 6).

Figure 6

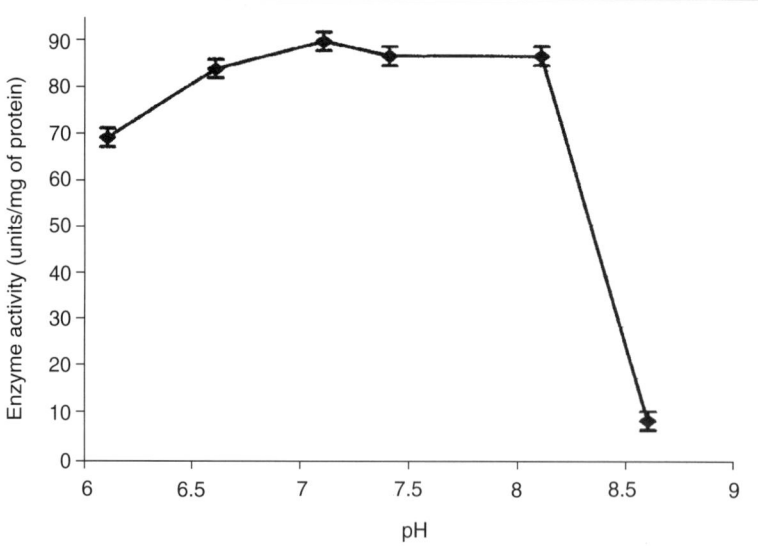

Effect of pH variation on diadenosine polyphosphate hydrolase activity

Figure 7

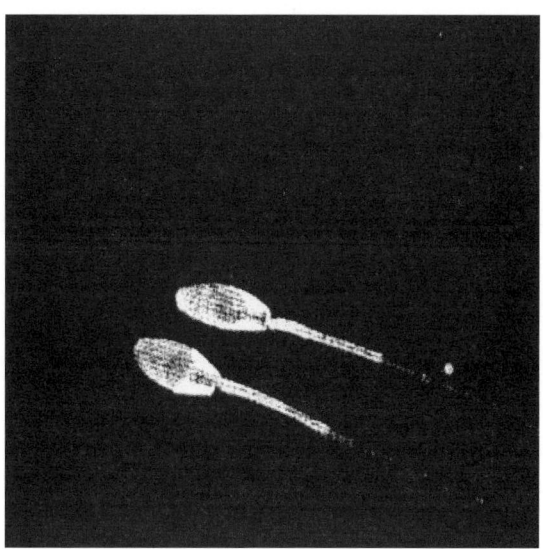

Immunofluorescent localization of A₁ adenosine receptors on horse sperm cells

The anti-A₁ receptor antibody (PC21) was a gift from the laboratory of R. Franco.

Maximal activity was obtained in the pH range 7.0–8.1, although the enzyme still presented a fairly high activity at pH 6.1.

The ecto-enzyme of the prostasome-like vesicles degraded diadenosine tetraphosphate and the main final product of degradation was adenosine (Table 1). ATP was not detected, meaning that the nucleotide was hydrolysed as soon as it was formed by the asymmetrical split of the substrate, i.e. ATP and AMP. The increase in ADP mirrored the disappearance of diadenosine tetraphosphate, indicating that the dephosphorylation of ADP is regulated in a different manner.

Horse sperm cells did not show any hydrolytic activity that was capable of degrading diadenosine compounds, but they did acquire such a capability after fusion with the seminal vesicles. It is plausible to suggest that the newly acquired enzymic activity may have somehow affected the horse sperm's physiology.

A₁ adenosine receptors are present on horse spermatozoa (Figure 7) and we have shown that stimulation of these receptors has a capacitative effect on sperm capacitation [28]. Therefore, we suggest that the addition of diadenosine tetraphosphate to a suspension of uncapacitated sperm might help the capacitative process, thereby increasing the fertilization potential of the sperm cells.

We thank Dr Mary Kerrigan for valuable linguistic suggestions.

References

1 Davis, B.K. (1973) Experientia **29**, 1484–1487
2 Breitbart, A. and Rubinstein, S. (1982) Arch. Androl. **9**, 147–157
3 Ronquist, G. and Brody, I. (1985) Biochem. Biophys. Acta **822**, 203–218
4 Agrawal, Y. and Vanha-Pertulla, T. (1986) Int. J. Biochem. **18**, 725–729
5 Fornes, M.W., Barbieri, A., Sosa, M.A. and Bertini, F. (1991) Andrologia **23**, 347–351

6   Kelly, R.W. (1991) Int. J. Androl. **14**, 242–247

7   Nilsson, B.O., Jin, M., Einarsson, B., Persson, B.E. and Ronquist, G. (1998) Prostate **35**, 178–184

8   Carlsson, L., Pahlson, C., Bergquist, M., Ronquist, G. and Stridsberg, M. (2000) Prostate **44**, 279–286

9   Carlsson, L., Lennartsson, L., Nilsson, S. and Ronquist, G. (2000) Eur. Urol. **38**, 468–474

10   Stegmayr, B. and Ronquist, G. (1982) Scand. J. Urol. Nephrol. **16**, 85–90

11   Davis, B.K. and Hungund, B.J. (1976) Biochem. Biophys. Res. Commun. **69**, 1004–1010

12   Agrawal, Y. and Vanha-Pertulla, T. (1987) J. Reprod. Fertil. **79**, 409–419

13   Ronquist, G., Nilsson, B.O. and Hejrten, S. (1990) Arch. Androl. **24**, 147–157

14   Laurell, C.B., Weiber, H., Ohlsson, K. and Rannevik, G. (1982) Clin. Chim. Acta **126**, 161–170

15   Mateo, J., Miras-Portugal, M.T. and Rotllan, P. (1997) Am. J. Physiol. **273**, C918–C927

16   Ronquist, G., Frithz, G. and Jansson, A. (1988) Urol. Int. **43**, 193–198

17   Ronquist, G. and Frithz, G. (1986) Acta Eur. Fertil. **17**, 273–276

18   Fabiani, R. and Ronquist, G. (1993) Clin. Chim. Acta **216**, 175–182

19   De Mester, L., Vanhof, G., Lambeir, A.M. and Scharpe, S. (1996) J. Immunol. Methods **189**, 99–105

20   Arienti, G., Carlini, E. and Palmerini, C.A. (1997) J. Membr. Biol. **155**, 89–94

21   Miras-Portugal, M.T., Gualix, J. and Pintor, J. (1998) FEBS Lett. **430**, 78–82

22   Kisselev, L.L., Justesen, J., Wolfson, A.D. and Frolova, L.Y. (1998) FEBS Lett. **427**, 157–163

23   McLennan, A.G. (1992) Ap4a and Other Dinucleoside Polyphosphates, CRC Press, Boca Raton, FL

24   Walker, J., Bossman, P., Lackey, B.R., Zimmermann, J.K., Dimmick, M.A. and Hilderman, R.H. (1993) Biochemistry **32**, 14009–14014

25   Rodriguez-Pascual, F., Cortes, R., Torres, M., Palacios, J.M. and Miras-Portugal, M.T. (1997) Neuroscience **77**, 247–255

26   Edgecombe, M., McLennan, A.G. and Fischer, M.J. (1996) Biochem. J. **314**, 687–693

27   Verspohl, E.J., Johanville, B., Kaiserling-Buddemeier, I., Schulter, H. and Hageman, J. (1999) J. Pharm. Pharmacol. **51**, 1175–1181

28   Allegrucci, C., Liguori, L. and Minelli, A. (2001) Biol. Reprod. **64**, 1653–1659

8

Immunomodulation in semen: the prostasome contribution

Rodney W. Kelly[1] and Elena Faccenda
Medical Research Council Human Reproductive Science Unit, University of Edinburgh Centre for Reproductive Biology, 37 Chalmers Street, Edinburgh EH3 9ET, U.K.

Introduction

Most invading organisms secrete immunomodulators to alter the immune responses of the host in such a way that their chances of survival are improved [1–4]. Spermatozoa are allogeneic invaders of the female tract and as such they are also likely to need protection against the female's immune defences, against attack from both the innate immune system and a chronic adaptive immune response. Many of the components of human seminal plasma, including prostasomes, are likely to fulfil this purpose.

An adaptive immune system that turned its response to spermatozoa would pose a threat to the continuation of the species. Although spermatozoa are allogeneic with reference to the female's immune system, the balance of evidence is that mature human spermatozoa do not express detectable amounts of HLA molecules on their surface [5]. Thus the immunological profile of the spermatozoa would not be high, and adaptive immune responses might be avoided up to the time that infection appeared in either the male or the female reproductive tract. The mammalian immune system has to distinguish both self from non-self and noxious from harmless and, in doing so, derives signals from pathogen-associated molecular patterns [6], with compounds such as lipopolysaccharide, which derive from bacterial cell walls, being used as evidence of infection and consequently 'danger'. In the case of spermatozoa, for example delivered into an infected female reproductive tract, the recipient immune system has all the signals present to mount a hostile response to the spermatozoa. Under these circumstances the evolutionary pressure to resolve the problem by adding immunosuppressive agents to semen is very strong. Once infection appeared, probably relatively late in evolution, the solution to the problem would have varied from species to species. In some species there may still be no problem, with no serious induction of an adaptive immune response, and this could account for the widely varying levels and types of immunomodulator found in the semen of different species.

The presence of infection in the male or female reproductive tracts will be related indirectly to sexual behaviour and numbers of partners. Thus species that are strictly monogamous or those with sexual intercourse controlled by the time of the year or the ovarian cycle will not have been subject to the intense evolutionary pressures to cope with spermatozoal delivery in or into fluids that represent a maelstrom of infection and counter-infection defences.

The Old World primates, because of multiple sexual encounters, may be particularly exposed to disease of the reproductive tract and thus have diverted large resources to providing the immunomodulatory agents found in semen. An example is the very high level of 19-hydroxyprostaglandins of the E series (PGEs) in semen from apes and Old World monkeys [7].

Two immunomodulatory activities may also have evolved in human seminal plasma: the acute inactivation of effector cells, such as monocytes, and an overriding bias imposed on antigen-presenting cells such that spermatozoal antigens presented to T-cells result in neutral or beneficial responses.

The major immunomodulatory agents present in human seminal plasma are the prostaglandins and the prostasomes [8]. Although PGEs may modulate both the innate and the adaptive immune systems, the prostasomes appear to have major defensive properties related to the innate system (Figure 1). The prostasomes appear to share immunosuppressive activity with other particulate organelles and these parallel systems suggest that the control of immune responses by particles is a widespread phenomenon.

Effects on the innate immune system

There is now evidence that PGE, a major component of human semen, has potent effects on antigen presentation and this suggests that the function of the high prostaglandin concentrations in semen is to protect the spermatozoa. However,

Figure 1

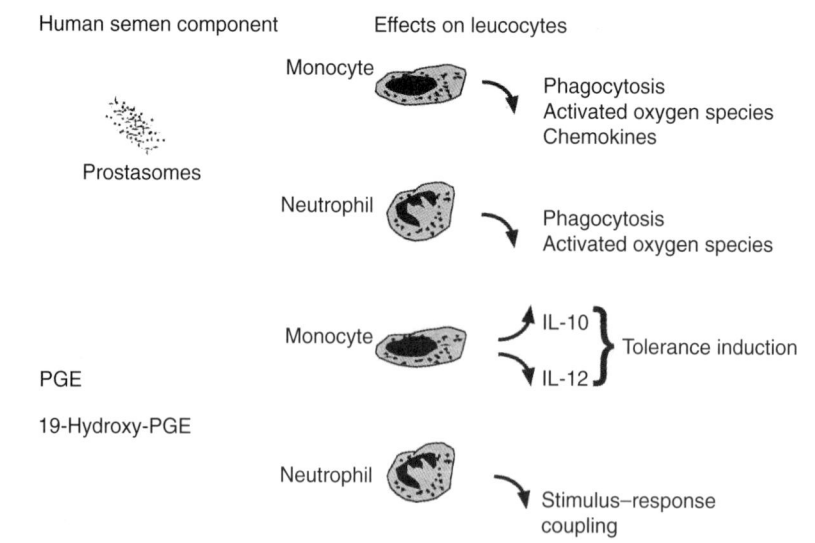

The effects of the major immunomodulators in human semen

The prostasomes have down-regulatory effects on phagocytic cells, whereas the prostaglandins affect cytokine release in such a way that antigen presentation will result in a Th-2 (T-cells that respond preferentially to antigens with a humoral or antibody response) or a regulatory T-cell response.

PGE also has important actions in down-regulating the main players in the innate immune system: monocytes, neutrophils and natural-killer (NK) cells. Monocyte activity is inhibited by three major agents, interleukin (IL)-10, transforming growth factor β (TGFβ) and PGE [9,10]. The mechanisms are complicated because of interactions. Thus IL-10 inhibits cyclo-oxygenase 2 production and consequently PGE release, and PGE has a profound effect in stimulating IL-10 [11]. A major effect of PGE is to down-regulate the neutrophil, as described in both early studies [12] and a later review [13]. Thus the elevation of intracellular cAMP in neutrophils will impair their activation and consequently their role in the innate immune system.

There have been several reports of PGE inhibiting the lytic activity of NK cells [14–16] and the action of the 19-hydroxy-PGE in semen appears to be a major contributor to NK cell inactivation [17].

Inactivation of monocytes and neutrophils by human seminal plasma will also be induced by the particulate content, the prostasomes. The prostasome, as a sub-micron organelle, may mimic organelles produced by cells, such as neutrophils, either as an autodefence mechanism [18] or after these cells have undergone apoptosis [19]. Prostasomes have been shown to inhibit T-cell prolif-eration [8] and this is likely to be due to an action on accessory cells such as monocytes in peripheral-blood mononuclear cell preparations. In addition, it can be shown that prostasomes inhibit a monocytic cell line (differentiated U937 cells) such that pre-incubation with prostasomes reduces the release of chemokines such as IL-8 and monocyte chemotactic protein-1 (Figure 2). The effect of prostasomes on another monocyte cell line (THP-1) is to reduce the phagocytosis of fluorescent beads by either macrophages or neutrophils [20]. Consistent with the inactivating action of prostasomes on neutrophils, prostasome preparations will inhibit oxygen free radical production by neutrophils [20]. In these ways prostasomes can prevent activation of macrophages and the phagocytic function of neutrophils and macrophages that might occur in the female tract after sperm deposition.

Effects on the adaptive immune response

There is a major need to protect the spermatozoa because an adaptive immune response against them would lead to anaphylaxis, the fulminating immune response that can be life threatening. The effect of PGE on monocytes is to stimulate IL-10 [11] and inhibit IL-12 [10] and this is the major effect of the human seminal-plasma prostaglandins. This critical cytokine switch is also effected by 19-hydroxy-PGE [21]. The effect of such a switch will mean that antigen is presented in the absence of IL-12 and the presence of IL-10 which will ensure either a Th-2 (T-cells that respond preferentially to antigens with a humoral or antibody response) or a regulatory T-cell (Tr) response [22]. The regulatory T-cell is of particular interest because it is one of the main cell types involved in oral tolerance, i.e. the ability of the gastrointestinal tract to recognize certain antigens, such as those from food, as harmless. The regulatory T-cell of the gut is induced by IL-10 and, when presented with its cognate antigen, responds by synthesis of further IL-10 together with IL-5, but very little IL-2, IL-4 or

Figure 2

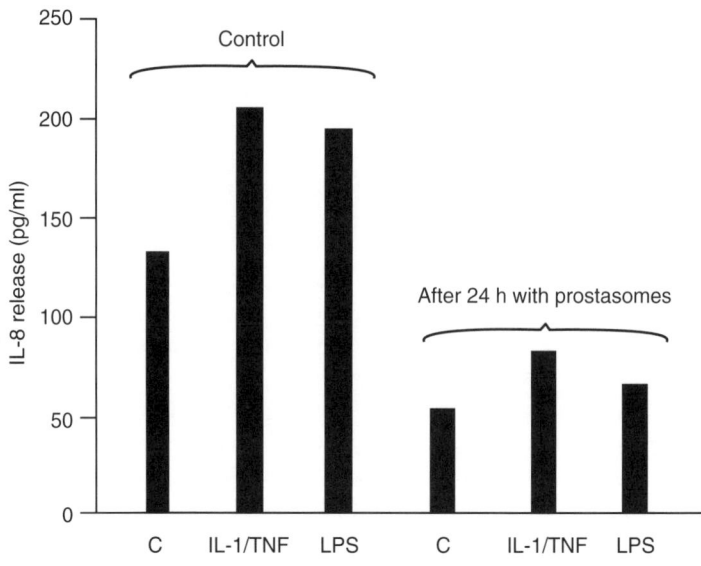

Prostasomes incubated with a monocyte cell line (U937) cause the reduction of IL-8 secretion, as measured by ELISA, in unstimulated and stimulated cells

C, control; IL-1/TNF, cells treated with 1 ng/ml IL-1 and 1 ng/ml tumour necrosis factor α; LPS, cells treated with 100 ng/ml lipopolysaccharide.

interferon-γ. This distinguishes the Tr1 cell from other regulatory cells of the gut that have been known to mainly produce TGFβ. Although prostaglandin could affect the generation of regulatory T-cells, this would appear to need the intermediate synthesis of IL-10. What is clear is that agents that inhibit prostaglandin synthesis, such as indomethacin, can break tolerance [23,24], and the adverse effects of non-steroidal anti-inflammatory drugs in inducing ulceration of the gut are well known, as is the ability of synthetic analogues of PGE, such as Misoprostol (Cytotec), to counter ulceration.

A more recent concept of antigen presentation is that dendritic cells derive a 'third' signal at the mucosal surface and that this signal controls the polarization of dendritic cells so that such a cell encountering PGE at the mucosal surface, described as a type-3 dendritic cell, has a reduced capacity for IL-12 synthesis. These antigen-presenting cells, when priming T-cells at the lymph node, will not generate T-helper-1 cells, although the T-cell phenotype produced is not so clear [25,26]. Currently the effects of prostaglandins on maturing dendritic cells are uncertain. The effects of prostaglandin on dendritic-cell maturation are still difficult to interpret, partly due to measurements of the p40 subunit of IL-12 [26a] being used incorrectly as a measure of IL-12 [27,28] and partly because of synthesis of prostaglandin by the dendritic cell itself.

Although prostaglandins can have a major effect on antigen-presenting cells at the mucosal surface, there is the possibility that prostaglandin or prostasomes could be taken by lymph drainage to the lymph node. Since, in animal models, it has been shown that cells and cell debris can reach the lymph node [29], it is certain that humoral agents (prostaglandin) or sub-micron particles (prostasomes) will also reach this organ, although the actions of these agents within the lymph node are currently unknown.

Parallel particulate systems

The presence of particles released in other activated cell systems might provide clues as to the purpose of the prostasomes. There is now growing evidence that a multitude of secreted vesicles arise from cells as varied as reticulocytes and antigen-presenting cells. These vesicles have been termed exosomes [30], and they have lipid bilayer membranes and are 50–100 nM in diameter. A further distinguishing feature of almost all types of exosomes is that they display the glycosylphosphatidylinositol (GPI)-linked complement inhibitor CD59 on their surface, as do the prostasomes. In this chapter these systems are examined and the presence of particulate material from the early trophoblast is identified as further evidence of common mechanisms used in benefitting both spermatozoa in semen and trophoblasts in early pregnancy.

Communication between cells through particulates occurs when neutrophils are activated [18]. Neutrophils activated with agents such as N-formylmethionyl-leucylphenylalanine ('fMLP') bud off small particles (approx. 100 nM in diameter) with bilayer membranes. These organelles, which have been termed ectosomes, like prostasomes display inhibitors of complement, and the ratio of the GPI-linked inhibitors differs on the ectosome surface from that seen on the surface of the original neutrophils. Thus CD59 is displayed prominently on the ectosome whereas CD55 and CD46 are almost absent [18]. The ectosomes also contain the protease elastase (which is normally found in the azurophilic granules of the neutrophil), which in the ectosome is found associated with complement receptor-1 (CR1), which is present on the surface of the organelle. Since the ectosomes are produced from activated neutrophils within minutes, it is unlikely that they are the result of apoptotic bodies being released from the cells. Ectosomes are released after migration of neutrophils into tissue and these organelles can be recovered from skin-blister fluid [18].

Another organelle that is formed with an enhanced concentration of the GPI-linked proteins CD55 and CD59 on its surface is the exosome, formed by maturing erythrocytes [31]. Other exosomes formed by dendritic cells and which carry MHC-associated antigen fragments also have CD59 on their surface [32]. The release of these exosomes is seen as a mechanism of up-regulating the immune system by increased antigen presentation but clearly such vesicles do not possess the necessary co-stimulatory molecules to match the professional antigen-presenting cell, and the effects of these vesicles might be to induce anergic responses. Although little is understood at present about antigen-presenting-cell-derived exosomes that carry processed antigens, any paternal antigen delivered in seminal-plasma prostasomes could explain improved placentation in long-term compared with short-term co-habitation prior to conception [33].

The ectosomes derived from activated neutrophils are unlikely to be due to apoptosis since they are produced within a few minutes of activation [18]. However, apoptotic neutrophils do produce small particles that are phagocytosed readily by macrophages. The result of this ingestion is down-regulation of the macrophage, shown by reduced secretion of IL-1β, IL-8, tumour necrosis factor α, IL-10 and thromboxane B2. This is accompanied by an increase in the secretion of TGFβ , PGE and platelet-activating factor [19]. One reason for this inactivation of macrophages by apoptotic cells is that it may represent the repair phase of an inflammatory reaction. The mechanism by which pro-inflammatory cytokines are decreased is not likely to be mediated by IL-10 since anti-IL-10 antibody failed to restore the cytokine release from treated macrophages. The immunosuppression of macrophages by apoptotic debris has parallels with the actions of prostasomes on phagocytosis and oxygen free radical production by macrophages [20].

The mechanisms of immunomodulation in human seminal plasma bear a striking resemblance to immunosuppressive activity shown by the trophoblast. Thus trophoblast antigens are also seen in human seminal plasma [34,35] and the trophoblast is a source of high levels of prostaglandin. The distribution of synthetic and catabolic enzymes for the prostaglandins shows that the fetal side of the trophoblast villi is protected from high prostaglandin levels, suggesting an outwards secretion of compounds such as PGE. After the culture of first-trimester villous trophoblasts for several days, on centrifugation at 100 000 g the medium gave a glassy pellet, similar to that seen after centrifugation of human seminal plasma. This pellet displayed antigen markers in common with prostasomes, suggesting a degree of similarity between particulate matter derived from trophoblasts and the prostasomes present in human seminal plasma. This is a further indication of mechanisms that are common between semen and trophoblasts.

Relevance of prostasomes to immune surveillance

The immunological properties of prostasomes suggest that they contribute to the protection of spermatozoa against cytotoxic and phagocytic agents of the immune system. The association of complement inhibitors with prostasomes has been described [36], but prostasomes are also active in the down-regulation of phagocytic cells and this property is shared with a variety of other particulate systems. Thus it is likely that the immunosuppressive activity of agents such as prostasomes and prostaglandins may contribute to the transmission of sexually transmitted disease, because mechanisms that provide protection for spermatozoa will also protect bacteria and viruses. Is it coincidental that some of the major pathologies of the reproductive tract, cervical and prostate carcinomas, arise in tissues that are in close contact with seminal plasma or its constituents? Although carcinoma of the cervix is related to exposure to semen, seminal fluid will carry both the papilloma virus and the immunomodulators, and the contribution of these two elements to initial infection and disease progression is difficult to establish. In the male tract the incidence of carcinoma of the prostate is not matched by disease in the other accessory reproductive organs, such as the seminal vesicles, and the different derivation of this tissue (from the mesoderm and Wolfian duct, as opposed to the urogenital sinus for the prostate) may be a major consideration in the initial cell transformation and subsequent course of the disease. An alternative explanation is that Nature may have separated immunosuppressive activity between accessory glands so that a potent combination is only achieved at the time of ejaculation. If this were so then leakage of agents such as prostaglandins from the seminal vesicle would reach the prostate, but since prostasomes are released by exocytosis or diacytosis, potentially more controllable mechanisms of release, prostasomes are unlikely to reach the seminal vesicle.

In considering the risk of infection carried in human seminal plasma, account has to be taken of anti-microbial activity present in human semen. Viral and bacterial infection in reproductive tracts is countered by the high concentrations of innate immune defences, such as lysosyme, lactoferrin, secretory leucocyte protease inhibitor and defensin-like molecules such as human cationic antimicrobial protein-18 [37]. Nevertheless, several aspects of immune defence are compromised. Thus T-cell priming, T-cell replication, NK-cell killing, and monocyte and macrophage function are all affected and, perhaps more importantly for the progression of transformed cells into carcinoma, immune surveillance is restricted.

Summary

There is an extensive range of compounds in human seminal plasma that have effects on the immune system. The adaptive immune system is probably modulated by the prostaglandins to ensure that a neutral or benign population of T-cells results from presentation of spermatozoal antigens. The more acute responses of the system will be repressed by a combination of prostasomes and prostaglandins. The prostasomes are not only inhibitors of the complement cascade, they also inhibit neutrophilic and monocytic functions. In this way they resemble the small vesicles released from other cell systems that have been shown to be immunosuppressive.

References

1   Maizels, R.M., Bundy, D.A.P., Selkirk, M.E., Smith, D.F. and Anderson, R.M. (1993) Nature (London) **365**, 797–805
2   Belley, A. and Chadee, K. (1995) Parasitol. Today **11**, 327–334
3   Beverley, S.M. (1996) Cell **87**, 787–789
4   Liu, L.X., Serhan, C.N. and Weller, P.F. (1990) J. Exp. Med. **172**, 993–996
5   Hutter, H. and Dohr, G. (1998) J. Reprod. Immunol. **38**, 101–122
6   Medzhitov, R. and Janeway, Jr, C.A. (1997) Cell **91**, 295–298
7   Kelly, R.W., Taylor, P.L., Hearn, J.P., Short, R.V., Martin, D.E. and Marston, J.H. (1976) Nature (London) **260**, 544–555
8   Kelly, R.W., Holland, P., Skibinski, G., Harrison, C., McMillan, L., Hargreave, T.B. and James, K. (1991) Clin. Exp. Immunol. **86**, 550–556
9   Bogdan, C., Paik, J., Vodovotz, Y. and Nathan, C. (1992) J. Biol. Chem. **267**, 23301–23308
10   Kraan, T.C.T.M.V., Boeije, L.C.M., Smeenk, R.J.T., Wijdenes, J. and Aarden, L.A. (1995) J. Exp. Med. **181**, 775–779
11   Strassmann, G., Patil Koota, V., Finkelman, F., Fong, M. and Kambayashi, T. (1994) J. Exp. Med. **180**, 2365–2370
12   Weissmann, G., Dukor, P. and Zurier, R.B. (1971) Nat. New Biol. **231**, 131–135
13   Reibman, J., Haines, K. and Weissmann, G. (1990) Curr. Topics Membr. Tr. **35**, 399–424
14   Leung, K.H. and Koren, H.S. (1982) J. Immunol. **129**, 1742–1747

15   Leung, K.H. (1989) Cell. Immunol. **123**, 384–395
16   Quayle, A.J., Kelly, R.W., Hargreave, T.B. and James, K. (1989) Clin. Exp. Immunol. **75**, 387–391
17   Tarter, T.H., Cunningham-Rundles, S. and Koide, S.S. (1986) J. Immunol. **136**, 2862–2867
18   Hess, C., Sadallah, S., Hefti, A., Landmann, R. and Schifferli, J.A. (1999) J. Immunol. **163**, 4564–4573
19   Fadok, V.A., Bratton, D.L., Konowal, A., Freed, P.W., Westcott, J.Y. and Henson, P.M. (1998) J. Clin. Invest. **101**, 890–898
20   Skibinski, G., Kelly, R.W., Harkiss, D. and James, K. (1992) Am. J. Reprod. Immunol. **28**, 97–103
21   Kelly, R.W., Carr, G.C. and Critchley, H.O.D. (1997) Hum. Reprod. **12**, 677–681
22   Groux, H., O'Garra, A., Bigler, M., Rouleau, M., Antonenko, S., deVries, J.E. and Roncarolo, M.G. (1997) Nature (London) **389**, 737–742
23   Scheuer, W.V., Hobbs, M.V. and Weigle, W.O. (1987) Cell. Immunol. **104**, 409–418
24   Louis, E., Franchimont, D., Deprez, M., Lamproye, A., Schaaf, N., Mahieu, P. and Belaiche, J. (1996) Int. Arch. Allergy Immunol. **109**, 21–26
25   Kalinski, P., Hilkens, C.M.U., Wierenga, E.A. and Kapsenberg, M.L. (1999) Immunol. Today **20**, 561–567
26   Kalinski, P., Schuitemaker, J.H.N., Hilkens, C.M.U., Wierenga, E.A. and Kapsenberg, M.L. (1999) J. Immunol. **162**, 3231–3236
26a   Kalinski, P., Vieira, P.L., Schuitemaker, J.H., de Jang, E.C. and Kapsenberg, M.L. (2001) Blood **97**, 3466–3469
27   Kalinski, P., Schuitemaker, J. and Kapsenberg, M.L. (1998) J. Leukocyte Biol. H2
28   Rieser, C., Böck, G., Klocker, H., Bartsch, G. and Thurnher, M. (1997) J. Exp. Med. **186**, 1603–1608
29   Spira, A.I., Marx, P.A., Patterson, B.K., Mahoney, J., Koup, R.A., Wolinsky, S.M. and Ho, D.D. (1996) J. Exp. Med. **183**, 215–225
30   Denzer, K., Kleijmeer, M.J., Heijnen, H.F., Stoorvogel, W. and Geuze, H.J. (2000) J. Cell Sci. **113**, 3365–3374
31   Rabesandratana, H., Toutant, J.P., Reggio, H. and Vidal, M. (1998) Blood **91**, 2573–2580
32   Clayton, A., Court, J., Navabi, H., Adams, M., Mason, M.D., Hobot, J.A., Newman, G.R. and Jasani, B. (2001) J. Immunol. Methods **247**, 163–174
33   Robillard, P.Y., Hulsey, T.C., Périanin, J., Janky, E., Miri, E.H. and Papiernik, E. (1994) Lancet **344**, 973–975
34   Thaler, C.J., Critser, J.K., McIntrye, J.A. and Faulk, W.P. (1989) Fertil. Steril. **52**, 463–468
35   Kajino, T., Torry, D.S., McIntyre, J.A. and Faulk, W.P. (1988) Am. J. Reprod. Immunol. **17**, 91–95
36   Rooney, I.A., Atkinson, J.P., Krul, E.S., Schonfeld, G., Polakoski, K., Saffitz, J.E. and Morgan, B.P. (1993) J. Exp. Med. **177**, 1409–1420
37   Malm, J., Sorensen, O., Persson, T., Frohm-Nilsson, M., Johansson, B., Bjartell, A., Lilja, H., Stahle-Backdahl, M., Borregaard, N. and Egesten, A. (2000) Infect. Immun. **68**, 4297–4302

9

Transfer of prostasomal CD59 to CD59-deficient erythrocytes results in protection against complement-mediated haemolysis

Adil A. Babiker*[1], Gunnar Ronquist*, Ulf R. Nilsson† and Bo Nilsson†
*Department of Medical Sciences, Clinical Chemistry, University Hospital, SE-751 85 Uppsala, Sweden, and †Department of Oncology, Radiology and Clinical Immunology, University of Uppsala, Uppsala, Sweden

Introduction

The complement system provides the organism with a natural array of defence against infectious agents in the genital tract [1,2]. Different pathways (the classical, the lectin and the alternative pathways) activate the complement system. All pathways end in the formation of the membrane-attack complex (C5b-9) [3,4], which causes cell lysis and activates leucocytes in inflammation [5]. Complement activation on the cell surface is cotrolled by a group of protein regulators that limit spontaneous activation of the complement system. This group includes the fluid-phase factor H, the membrane-bound complement receptor-1 (CR1), decay-accelerating factor (DAF), membrane cofactor protein (MCP) and CD59 [6,7]. CD59, also known as membrane inhibitor of reactive lysis (MIRL), prevents the formation of the membrane-attack complex on the host cells by interfering with the C8 and C9 interaction and C9 polymerization. CD59 is a glycosylphosphatidyl-inositol (GPI)-anchored complement-modulating glycoprotein of 18–21 kDa and is expressed on the surface of many cells, including erythrocytes [8].

Rabbit erythrocytes (REs) lack regulators of convertase in the alternative pathway in human serum [6,9,10] and also lack a functional CD59 against human complement. The erythrocytes in patients with paroxysmal nocturnal haemoglobinuria (PNH) also mainly lack CD59 and DAF [11]. PNH is an acquired uncommon chronic clonal disorder characterized by increased susceptibility of erythrocytes to complement-mediated lysis due to the absence of two regulators of complement activators that are responsible for the protection of cells against complement attack [12,13]. Those proteins are DAF [14] and CD59 [5,8,12,13,15,16]. Some observations suggested that CD59 is the more important deficient complement-modulating protein in the pathogenesis of PNH. After lysis of some of the erythrocytes in patients with PNH, haemoglobin is passed in the urine. The erythrocytes susceptible to lysis are classified as PNH I, II or III, with PNH III being the most liable to lysis. The intensity of the clinical process in PNH is different from patient to patient and is related to the size of the PNH III

[1]To whom correspondence should be addressed (e-mail adil.babiker@medsci.uu.se).

population. If PNH III erythrocytes comprise less than 20% of the population, then haemolysis will be mild or undetectable. If they range from 20 to 50%, then episodes of haemoglobinuria will occur. If they are more than 50% then constant haemoglobinuria will occur.

Prostasomes are prostate-derived secretory granules or vesicles with a mean diameter of 150 nm (range 40–500 nm) that are present in human semen [17,18]. They have a complement-regulatory effect [19] mainly due to their CD59 content, which is present in seminal plasma at a concentration of 20–30 µg/ml [20,21]. Rooney et al. [21] showed that the GPI-anchored complement-regulatory proteins of seminal prostasomes could be transferred to spermatozoa and guinea-pig erythrocytes. The aim of the work presented here was to study the possibility of transferring prostasomal CD59 to CD59-deficient red blood cells of two different species, so that these cells would be able to resist lysis by the human complement system. We used REs and human erythrocytes obtained from patients with PNH for incubation experiments with prostasomes.

Materials and methods

Phosphatidlyinositol-specific phospholipase C (PIPLC) from *Bacillus cereus* was purchased from Calbiochem-Novabiochem (Darmstadt, Gemany). Rat anti-human CD59 YTH53.1 monoclonal antibody, F(ab′)$_2$ rabbit anti-rat IgG-FITC and rat anti-human glycophorin A YTH89.1 IgG2b were purchased from Serotec (Oxford, U.K.).

Buffers and preparation of prostasomes

Gelatin was dissolved (0.1%, v/v) in veronal-buffered saline (VB^{2-}) to make gelatin-VB^{2-} (GVB^{2-}). MgCl$_2$ (200 mM) and EGTA were added to GVB$_2$ to make GVB-Mg EGTA buffer [21a]. The pH was adjusted to 7.4. EDTA buffer: EDTA was prepared as a stock solution (200 mM). VB was added and diluted to make a buffer solution containing 10 mM EDTA, pH 7.5.

Prostasomes were isolated from human semen and prepared as described by Ronquist and Brody [17]. Prostasome concentrations were expressed in mg of protein/ml. Prostasomes (2 mg/ml) as stock solutions were heat-treated at 60, 70, 80, 90 and 100°C for 30 min. A 100 µg/ml suspension was made of each solution.

As described by Fabiani and Ronquist [22], PIPLC was used to treat the prostasomes in order to break the GPI anchor and release CD59. The PIPLC concentration was 2 units/ml (in Tris/HCl buffer).

Haemolysis tests

REs were washed three times with GVB-Mg EGTA buffer. An RE suspension was prepared (10^8 cells/ml). PNH erythrocyte samples were collected from three patients who had already been diagnosed as having PNH. Cells were washed three times in GVB-Mg EGTA buffer. A PNH cell suspension was prepared (10^8 cells/ml). Normal human erythrocytes were subjected to the same washing procedure. A normal erythrocyte suspension was prepared (10^8 cells/ml).

To test the alternative pathway of the complement system, REs were incubated with serum at 37°C for 30 min in a shaking water bath. The incubation was terminated by adding cold EDTA buffer. Centrifugation followed and the absorbance of the supernatant was monitored in a spectrophotometer at 405 nm as a measure of haemolysis.

Preincubation of erythrocytes with prostasomes

An RE suspension was preincubated with prostasomes for 2 h. REs were washed three times with GVB-Mg EGTA buffer to remove the prostasomes. The REs were then incubated with serum at 37°C for 30 min until termination as described above. REs preincubated in buffer lacking prostasomes were used as controls.

A PNH cell suspension was also preincubated with prostasomes for 2 h. PNH cells were washed three times with GVB-Mg EGTA buffer followed by incubation with serum at 37°C for 30 min until termination. PNH erythrocytes and normal erythrocytes preincubated in buffer lacking prostasomes were used as controls.

A heated prostasome suspension was preincubated with REs. REs were washed three times with GVB-Mg EGTA buffer to remove the prostasomes. Unheated prostasomes preincubated with REs were used as a positive control and REs preincubated in buffer lacking prostasomes were used as a negative control. REs were incubated with serum at 37°C for 30 min until termination.

Preincubation of REs with prostasomes treated with PIPLC

Equal volumes of each of the supernatants and the PIPLC-treated prostasomes were preincubated with the RE suspension. The REs were washed three times with GVB-Mg EGTA buffer to remove the prostasomes. The positive control was REs preincubated with untreated prostasomes and the negative control was REs preincubated in buffer lacking any type of prostasomes. REs were incubated with human serum at 37°C for 30 min until termination.

PIPLC treatment of REs and PNH cells preincubated with prostasomes

An RE suspension was preincubated with the prostasome suspension. REs were washed three times with GVB-Mg EGTA buffer to remove the prostasomes, followed by incubation with PIPLC dissolved in GVB-Mg EGTA buffer (2 units/ml).

Antibody suppression of CD59 transfer

Rat anti-CD59 antibodies were preincubated with prostasomes for 1 h followed by another preincubation of the prostasomes for 2 h with a PNH erythrocyte suspension. PNH erythrocytes were washed three times with GVB-Mg EGTA buffer and incubated with serum. As controls, prostasomes preincubated with anti-glycophorin A antibodies and erythrocytes preincubated with prostasomes not previously incubated with the anti-CD59 antibodies were used. Incubation in serum continued for 30 min until termination as described above.

Flow cytometry

This assay was used to determine whether CD59 present on prostasomes had the capacity to be transferred and bind to the REs and PNH erythrocytes. REs and PNH erythrocyte suspensions preincubated with prostasomes were washed three times with GVB-Mg EGTA buffer, followed by another preincubation with rat anti-human CD59 antibodies for 30 min. Cells were washed three times with GVB-Mg EGTA buffer to remove unbound antibodies. Erythrocytes were incubated with F(ab')$_2$ rabbit anti-rat IgG conjugated to FITC. Cells were then washed three times with GVB-Mg EGTA buffer to remove unbound antibodies, before analysis using FACSCalibur® (Becton Dickinson).

Results

Anti-haemolytic effect of prostasomes on REs

REs preincubated with prostasomes acquired a dose-dependent resistance to complement-mediated lysis during the subsequent incubation with serum. The control erythrocytes preincubated in buffer lacking prostasomes remained susceptible to lysis during the subsequent incubation with serum. A saturation point was obtained at a prostasome concentration of about 40 µg/ml with very little additional effect of prostasomes up to a protein concentration of 2 mg/ml.

Thermolability of prostasomes

Prostasomes lost their ability to protect REs from complement-mediated lysis when heated at 100°C for 30 min. The erythrocytes incubated with non-heated prostasomes gained maximum protection.

Effect of pretreatment with PIPLC on prostasomes

REs preincubated with prostasomes already pretreated with PIPLC gained protection against subsequent complement-mediated lysis to a lesser extent than erythrocytes preincubated with prostasomes without this pretreatment. REs preincubated with the supernatant that had been obtained after the PIPLC pretreatment of prostasomes were also protected to some degree. REs preincubated with prostasomes and then treated with PIPLC lost most of the protection already gained.

Prostasome effect on PNH erythrocytes

PNH erythrocytes preincubated with prostasomes acquired resistance to complement-mediated lysis nearly to the level of normal erythrocytes whereas those preincubated with buffer lacking prostasomes remained susceptible to lysis.

Effect of anti-CD59 antibodies on prostasome-mediated protection

The PNH cells preincubated with prostasomes already pretreated with anti-CD59 antibodies showed less resistance to subsequent complement-mediated lysis than those preincubated with non-manipulated prostasomes.

Flow-cytometric findings

The flow-cytometric analysis showed that CD59 was transferred from prostasomes to REs during preincubation. The amount of CD59 acquired by REs was less when preincubated with heated prostasomes. The PNH erythrocytes acquired CD59 from the prostasomes in the same fashion.

Conclusion

Prostasomes have several biological activities but their physiological function is still unclear. They have immunosuppressive properties [19]. CD59 of seminal plasma and prostasomes could be acquired by sperm and some other cells. The results obtained from this study show that erythrocytes lacking CD59, and therefore susceptible to complement-mediated lysis, acquired resistance to lysis after preincubation with purified seminal prostasomes. This was found to be due to the acquisition of human CD59 from prostasomes.

The preincubation of PNH erythrocytes with prostasomes led to what could be described as a normalization of the erythrocytes with respect to resistance against complement attack in nearly the same way as normal human erythrocytes. The mechanism of transfer of CD59 from prostasomes to other cells is still unclear.

Heating was found to be unable to completely inhibit the protective effect of prostasomes, except at 100°C, which was probably due to the denaturation of proteins. This resistance to heat may be due to the unique membrane architecture of prostasomes with a very high degree of molecular ordering [23]. The action of PIPLC on the GPI anchor and its recovery in the supernatant suggested that the decrease in the complement-regulatory efficacy of prostasomes after PIPLC treatment was due to the release of CD59 into the medium after preparative ultracentrifugation. The acquired resistance of REs preincubated with prostasomes to complement-mediated lysis was also affected by PIPLC treatment of the preincubated erythrocytes, most probably due to its action on the GPI anchor of CD59. This effect was not applicable to human erythrocytes.

The data obtained with prostasomal CD59 support the idea that transfer of CD59 from prostasomes to autologous or allogeneic cells can protect these cells from lysis elicited by the membrane-attack complex (C5b-9). A similar transfer mechanism may be operative in the genital tract by which autologous and allogeneic cells are protected against complement attack.

*This article has been adapted from Babiker, A.A., Ronquist, G., Nilsson, U.R. and Nilsson, B. (2002) Transfer of prostasomal CD59 to CD59-deficient red blood cells results in protection against complement-mediated hemolysis. Am. J. Reprod. Immunol. **47**, 183–192, with permission from Blackwell Science Ltd.*

References

1   Jin, M., Larsson, A. and Nilsson, B.O. (1991) Am. J. Reprod. Immunol. **26**, 53–57
2   Hasty, L.A., Lambris, J.D., Lessey, B.A., Pruksananonda, K. and Lyttle, C.R. (1994) Am. J. Obstet. Gynecol. **170**, 168–175
3   Morgan, B.P. (2000) Methods Mol. Biol. **150**, 1–13
4   Muller-Eberhard, H.J. and Schreiber, R.D. (1980) Adv. Immunol. **29**, 1–53

5   Davies, A., Simmons, D.L., Hale, G., Harrison, R.A., Tight, H., Lachmann, P.J. and Waldmann, H. (1989) J. Exp. Med. **170**, 637–654

6   Fearon, D.T. and Austen, K.F. (1977) Proc. Natl. Acad. Sci. U.S.A. **74**, 1683–1688

7   Schreiber, R.D., Pangburn, M.K., Lesavre, P.H. and Muller-Eberhard, H.J. (1978) Proc. Natl. Acad. Sci. U.S.A. **75**, 3948–3952

8   Sugita, Y., Nakano, Y. and Tomita, M. (1988) J. Biochem. (Tokyo) **104**, 633–637

9   Horstmann, R.D., Pangburn, M.K. and Muller-Eberhard, H.J. (1985) J. Immunol. **134**, 1101–1104

10   Fearon, D.T. and Austen, K.F. (1977) J. Exp. Med. **146**, 22–33

11   Yamashina, M., Ueda, E., Kinoshita, T., Takami, T., Ojima, A., Ono, H., Tanaka, H., Kondo, N., Orii, T. and Okada, N. (1990) N. Engl. J. Med. **323**, 1184–1188

12   Holguin, M.H., Wilcox, L.A., Bernshaw, N.J., Rosse, W.F. and Parker, C.J. (1989) J. Clin. Invest. **84**, 1387–1394

13   Holguin, M.H., Fredrick, L.R., Bernshaw, N.J., Wilcox, L.A. and Parker, C.J. (1989) J. Clin. Invest. **84**, 7–17

14   Nicholson-Weller, A., March, J.P., Rosenfeld, S.I. and Austen, K.F. (1983) Proc. Natl. Acad. Sci. U.S.A. **80**, 5066–5070

15   Zalman, L.S., Wood, L.M., Frank, M.M. and Muller-Eberhard, H.J. (1987) J. Exp. Med. **165**, 572–577

16   Okada, N., Harada, R., Taguchi, R. and Okada, H. (1989) Biochem. Biophys. Res. Commun. **164**, 468–473

17   Ronquist, G. and Brody, I. (1985) Biochim. Biophys. Acta **822**, 203–218

18   Nilsson, B.O., Jin, M., Einarsson, B., Persson, B.E. and Ronquist, G. (1998) Prostate **35**, 178–184

19   Kelly, R.W., Holland, P., Skibinski, G., Harrison, C., McMillan, L., Hargreave, T. and James, K. (1991) Clin. Exp. Immunol. **86**, 550–556

20   Rooney, I.A., Atkinson, J.P., Krul, E.S., Schonfeld, G., Polakoski, K., Saffitz, J.E. and Morgan, B.P. (1993) J. Exp. Med. **177**, 1409–1420

21   Rooney, I.A., Heuser, J.F. and Atkinson, J.P. (1996) J. Clin. Invest. **97**, 1675–1686

21a Mayer, M.M. (1961) in Experimental Immunochemistry, Chapter 4 (Kabat, E.A. and Mayer, M.M. eds), Thomas, Springfield, IL

22   Fabiani, R. and Ronquist, G. (1995) Prostate **27**, 95–101

23   Arvidson, G., Ronquist, G., Wikander, G. and Ojteg, A.C. (1989) Biochim. Biophys. Acta **984**, 167–173

10

Expression of chromogranins in prostasomes indicates neuroendocrine differentiation

Mats Stridsberg
Department of Medical Sciences, Clinical Chemistry, University Hospital, SE-751 85
Uppsala, Sweden

The neuroendocrine system

The diffuse neuroendocrine system consists of peptide-hormone-producing cells in different locations throughout the body [1]. Usually, the cells in the anterior lobe of the pituitary gland, the thyroid C-cells, the parathyroid glands, the adrenal medulla, the islets of Langerhans and the hormone-producing cells in the gastro-intestinal tract are included in the neuroendocrine system. A common feature for all of these tissues and cells is their ability to secrete bioactive amines. Based on these observations, the amine precursor uptake and decarboxylation (APUD) concept was introduced [2]. Neuroendocrine cells (earlier called APUD cells) can be detected histochemically with different silver-staining methods. All neuro-endocrine cells are capable of synthesizing and releasing amines, hormones (neuropeptides and regulatory peptides) and common protein markers, so-called neuroendocrine markers, which can be detected immunohistochemically. The most used markers are chromogranins and synaptophysin, which are described below.

Neuroendocrine cells in the prostate

Neuroendocrine cells have been detected in the human prostate by histochemical and immunohistochemical methods [3]. The cells are located in the epithelium of the acini and ducts of the prostate gland. During development chromogranin-positive cells can be found in the prostate early after gestation. These neuro-endocrine cells seem to have a neurogenic origin, distinct from the urogenital sinus-derived prostate cells [4].

Chromogranin A (CgA) has been used as both an immunohistochemical and a circulating marker for prostate-derived tumours. In one study of patients with advanced prostate cancer a correlation was found between the number of CgA-positive cells in the tumours and the plasma levels of CgA, but no such correlation was found for chromogranin B (CgB) [5]. However, in a follow-up study of these patients CgB plasma levels were increased, maybe indicating accumulation of CgB in malignant prostatic cells [6]. In another study it was shown that circulating CgA was increased in 32% of the prostate cancer cases and that CgA could be used to identify a subgroup of patients with advanced prostate cancer, who did not have elevated serum prostate-specific antigen levels [7]. These data suggest that neuroendocrine differentiation of the prostate is an important feature and may have significant implications for both normal prostatic development and functions, as well as tumour development and growth.

The chromogranin and secretogranin family of proteins

The chromogranin/secretogranin family consists of CgA, CgB (sometimes called secretogranin I), secretogranin II (SgII; sometimes called chromogranin C), secretogranin III (often called 1B1075), secretogranin IV (often called HISL-19) and secretogranin V (often called 7B2) [8]. CgA, originally called secretory protein I, was first isolated as a water-soluble protein present in chromaffin cells from bovine adrenal medulla [9–11]. Later CgB was also isolated from the adrenal medulla [12]. SgII was isolated from the anterior pituitary [13,14].

Location of chromogranins

Chromogranins are found in neuroendocrine cells throughout the body, forming the neuroendocrine system. Chromogranins are also located in the neuronal cells in the central and peripheral nervous systems. Neuroendocrine cells synthesize and release amines and hormones (neuropeptides and regulatory peptides). These are stored intracellularly in secretory granules and are released upon stimulation. Besides hormones, the granules contain one or more members of the chromogranin/secretogranin family of proteins.

Adrenal chromaffin granules contain about 40 different proteins and the chromogranins account for approx. 40% of the soluble proteins [12]. In adrenal chromaffin cells CgA and CgB are present in about equal amounts, whereas SgII is less abundant [15,16]. However, secretory granules from other neuroendocrine tissues show different relative amounts of chromogranins and secretogranins. For example, parathyroid cells and enterochromaffin cells in the stomach contain mostly CgA, and very little CgB [15,17]. On the other hand, prostasomes, a secretory product from the prostate, contain mostly CgB [18].

Protein structure

The members of the chromogranin/secretogranin family have some common structural features, which are more or less unique for these proteins. They are synthesized with an N-terminal signal peptide, which directs the chromogranins/secretogranins through the Golgi apparatus to the regulated secretory pathway. They are hydrophilic and highly acidic, with a high glutamic acid content, which gives them an isoelectric point of approx. 5 [10]. Chromogranins/secretogranins have multiple pairs of basic amino acids distributed along their molecules, but these are more abundant in the C-terminal regions. These sites are potential cleavage sites for production of biologically active peptides. The proteins are heat-stable and can resist high temperatures, even boiling, without denaturation [8,13]. They can bind calcium and other bivalent cations, which can induce conformational changes [19,20]. The proteins may be O-glycosylated, phosphorylated and sulphated [13,21].

In addition, CgA and CgB have a similar structure in the N-terminal region, where two cysteine residues, at amino acid positions 17 and 38 for CgA and 16 and 37 for CgB, form a disulphide bridge. This probably gives the three-dimensional structure of the N-terminal region a more rigid 'loop' formation than the rest of the molecule, which mostly displays a random coil structure. Furthermore, this N-terminal part is well conserved between different species, which indicates its biological importance. This structure is not seen in the secretogranins [8].

Post-translational modifications

Chromogranins are processed post-translationally. CgA isolated from bovine adrenal medulla exhibits five phosphorylation sites and two O-glycosylation sites [21]. Six of these post-translational modifications were located in regions with highly conserved amino acid sequences, indicating their importance for biological functions. This biological importance was shown for the anti-bacterial peptide chromacin (corresponding to human CgA residues 176–225), which was biologically inactive without glycosylation [22]. Post-translational modifications have also been identified in human CgA [23], some of which were found at positions that corresponded with the bovine sequence. However, this sample of human CgA was isolated from urine collected from a patient with a carcinoid tumour and so it is possible that the post-translational processing may be somewhat different from that in 'normal' human CgA.

CgA can also be sulphated. It has been shown that CgA isolated from the bovine parathyroid gland is sulphated on tyrosine residues, whereas CgA isolated from bovine adrenals is sulphated mainly on oligosaccharide residues. CgB from bovine adrenals, on the other hand, was found to be sulphated on tyrosine residues [24]. This shows that different neuroendocrine cells are capable of processing the chromogranins differently. It is therefore possible that the post-translational differences may affect both the metabolism and the biological functions of the chromogranins.

Specific processing

Peptide hormones and neuropeptides are synthesized as propeptides and processed specifically to biologically active peptides within the secretory granules before release. The best known example is the processing of proinsulin to active insulin and C-peptide. The cleavage is mediated to a large extent by the pro-hormone convertases PC1/3 and PC2, which are present almost exclusively in neuroendocrine tissue [25,26]. These enzymes act preferentially on the C-terminal side of sites with two basic amino acids, usually Lys-Arg or Arg-Arg, and sometimes also at single Arg residues. Chromogranins and secretogranins are co-stored with the hormones and neuropeptides within the secretory granules and, because they also have several pairs of basic amino acids, specific cleavage of chromogranins will probably take place. This pro-hormone convertase-dependent cleavage will generate specific chromogranin-related peptides that may have biological importance. Indeed, a recent study has shown that endocrine cells within the pancreatic islets process the CgA molecule differently, giving rise to different fragments of CgA [27].

Biological effects of chromogranins and chromogranin-related peptides

Intracellularly located chromogranins have been suggested to play a role in the formation, function and regulation of secretory granules and their contents [28]. As indicated above, chromogranins can also serve as precursors for generation of biologically active peptides. Several chromogranin-related peptides have been identified in biological tissue and fluids, and some of these have also been shown to have biological functions.

The biological functions of the chromogranin-related peptides are located in different parts of the molecules. The N-terminal part of CgA has vasodilatory functions, whereas the central and C-terminal parts of CgA have autocrine or paracrine inhibitory functions on hormonal release. The C-terminal region of CgB has anti-bacterial functions. The middle part of SgII (secretoneurin) influences the migration of monocytes.

Metabolism of chromogranins

Unspecific degradation of chromogranins starts both C- and N-terminally [29]. It has also been shown that CgA is filtered in the kidneys, taken up in the proximal tubuli cells and degraded in the lysosomal pathway [30]. However, due to its size, the intact CgA molecule is not filtered in normal kidneys, but requires either cleavage of the entire molecule or kidney dysfunction that affects glomerular filtration for it to be released into the urine [31,32]. Like other small proteins and peptides, fragments of CgA can be both filtered in the glomeruli and metabolized in the proximal tubules.

Clinical significance of chromogranins

At the first presentation of radioimmunoassay methods for measurements of CgA the clinical use was outlined [33]. Patients with tumours of neuroendocrine origin usually present with increased plasma levels of CgA. The neuroendocrine tumours are derived from the neuroendocrine cells and typical neuroendocrine tumours are carcinoid tumours, pheochromocytomas, neuroblastomas, small-cell lung cancers, hyperparathyroid adenomas, pituitary tumours, prostate cancers and pancreatic islet tumours, including multiple endocrine neoplasia (MEN) 1 and 2 syndromes. This also includes the different pancreatic islet cell syndromes, namely insulinomas, glucagonomas, somatostatinomas, Zollinger–Ellison syndrome, Verner–Morrison syndrome and non-functioning neuroendocrine tumours. The most useful assays for tumour detection have been those that measure the whole CgA molecule. Assays measuring specifically defined parts of the molecule usually have lower sensitivity in detecting patients with neuroendocrine tumours [34].

Synaptophysin and related proteins

Synaptophysin was first isolated from the brains of calf [35] and rat [36]. By immunocytochemical studies of nervous tissue, it has been shown that synaptophysin is located in the membranes of the small synaptic vesicles [37]. In neuroendocrine tissue, synaptophysin is found in secretory granules. Antibodies against synaptophysin immunostain almost all normal neuroendocrine tissue and neuroendocrine-derived tumours [38].

The synaptophysin gene, which covers about 20 kb of DNA and contains seven exons and six introns, is located on the X chromosome at Xp11.22–p11.23 [39]. In humans, the translated 2.5 kb mRNA encodes a 296-amino acid chain with four hydrophobic regions and five hydrophilic regions [40,41]. Based on the amino acid sequence, the molecular mass was estimated as 33 kDa [40,42], but SDS/PAGE revealed molecular masses of about 38–42 kDa [37,43]. This is probably due to the different degrees of N-glycosylation of synaptophysin in different tissues. Synaptophysin can also bind calcium [43] and can be phosphorylated at serine residues in a calcium-dependent manner [44]. The repeating hydrophobic and hydrophilic regions indicate that the synaptophysin molecule is membrane bound. Thus synaptophysin forms a membrane-bound hexameric homo-oligomer structure that resembles the structure of a channel-protein complex [41,45].

Synaptophysin has also been found in the 5-hydroxytryptamine (serotonin)-containing veiscles in platelets [46]. Proteins with similar structure have been identified in granulocytes, lymphocytes, monocytes and mast cells [46]. The name granulophysin was introduced and it was shown that these proteins share common antibody-binding epitopes with synaptophysin [47].

Neuroendocrine markers in prostasomes

Several potential biological functions of prostasomes have been described [48]. Among them, neuroendocrine properties of the prostasomes have been noted [18]. In that study [18] we demonstrated the presence of the general neuroendocrine markers CgA, CgB and synaptophysin in prostasomes (Table 1). We also showed the expression of two neuropeptides, neuropeptide Y and vasoactive intestinal polypeptide, which are usually found co-existing with noradrenaline and acetylcholine in the nervous system and in neuroendocrine cells. In the prostasomes we found a striking predominance of CgB over CgA, with almost 50 times more CgB. Usually CgA is more commonly expressed and found in higher concentrations than CgB. A similar predominance of CgB has not been found

Table 1

Neuroendocrine peptides and proteins	Concentration (pmol/g of protein)
Synaptophysin	190
CgA	50
CgB	2300
Chromogranin C	<8
Neuropeptide Y	3000
Vasoactive intestinal polypeptide	2500

Expression of neuroendocrine peptides and proteins in prostasomes
Data are from [18].

elsewhere. Expression of the synaptophysin-related protein granulophysin in prostasomes has been shown previously [49].

In neuroendocrine cells chromogranins and biologically active peptides are stored in hormonal granules, which usually have a size range of 100–500 nm. This size range is similar to that of prostasomes, which present with a diameter of 40–500 nm. Thus the histological appearance and the size of the prostasomes are very similar to that of secretory granules. Furthermore, the presence of the common neuroendocrine markers, i.e. chromogranins and synaptophysin, together with neuropeptides, supports the idea that prostasomes can be considered as specialized secretory granules.

Speculations

Several biological functions have been attributed to the prostasomes [48]. What additional functions would one expect to see if the prostasomes were considered as secretory vesicles? First, amines and neuropeptides could transmit endocrine or paracrine signals that enhance the functions of sperm, for example facilitating the capacitation process. Secondly, substances released from the prostasomes could affect the female genital tract in a favourable way (for the sperm). Finally, chromo-granins have several biological effects, including anti-microbiological effects, which could be of importance for sperm function and fertility.

To conclude, prostasomes are unique extracellularly located organelles with multiple functions. To look at them from the neuroendocrine point of view could further expand our knowledge of prostasomes and their biological significance.

This work was supported by grants from the Swedish Cancer Foundation.

References

1   Polak, J.M. and Bloom, S.R. (1979) J. Histochem. Cytochem. **27**, 1398–1400
2   Pearse, A.G.E. (1969) J. Histochem. Cytochem. **17**, 303–313
3   Abrahamsson, P.A. and di Sant'Agnese, P.A. (1993) J. Androl. **14**, 307–309
4   Aumuller, G., Leonhardt, M., Janssen, M., Konrad, L., Bjartell, A. and Abrahamsson, P.A. (1999) Urology **53**, 1041–1048
5   Angelsen, A., Syversen, U., Haugen, O.A., Stridsberg, M., Mjolnerod, O.K. and Waldum, H.L. (1997) Prostate **30**, 1–6
6   Angelsen, A., Syversen, U., Stridsberg, M., Haugen, O.A., Mjolnerod, O.K. and Waldum, H.L. (1997) Prostate **31**, 110–117

7   Deftos, L.J., Nakada, S., Burton, D.W., di Sant'Agnese, P.A., Cockett, A.T.K. and Abrahamsson, P.A. (1996) Urology **48**, 58–62

8   Huttner, W.B., Gerdes, H.H. and Rosa, P. (1991) Trends Biochem. Sci. **16**, 27–30

9   Banks, P. and Helle, K. (1965) Biochem. J. **97**, 40C–41C

10   Smith, A.D. and Winkler, H. (1967) Biochem. J. **103**, 483–492

11   Blaschko, H., Comline, R.S., Schneider, F.H., Silver, M. and Smith, A.D. (1967) Nature (London) **215**, 58–59

12   Fischer-Colbrie, R. and Frischenschlager, T. (1985) J. Neurochem. **44**, 1854–1861

13   Rosa, P., Hille, A., Lee, R.H.W., Zanini, A., Decamilli, P. and Huttner, W.B. (1985) J. Cell Biol. **101**, 1999–2011

14   Fischer Colbrie, R., Hagn, C., Kilpatrick, L. and Winkler, H. (1986) J. Neurochem. **47**, 318–321

15   Hagn, C., Schmid, K.W., Fischer Colbrie, R. and Winkler, H. (1986) Lab. Invest. **55**, 405–411

16   Schober, M., Fischer Colbrie, R., Schmid, K.W., Bussolati, G., O'Connor, D.T. and Winkler, H. (1987) Lab. Invest. **57**, 385–391

17   Buffa, R., Mare, P., Gini, A. and Salvadore, M. (1988) Basic Appl. Histochem. **32**, 471–484

18   Stridsberg, M., Fabiani, R., Lukinius, A. and Ronquist, G. (1996) Prostate **29**, 287–295

19   Yoo, S.H. (1992) Biochemistry **31**, 6134–6140

20   Yoo, S.H. (1993) Biochim. Biophys. Acta Mol. Cell. Res. **1179**, 239–246

21   Strub, J.M., Sorokine, O., Van Dorsselaer, A., Aunis, D. and Metz-Boutigue, M.H. (1997) J. Biol. Chem. **272**, 11928–11936

22   Strub, J.M., Goumon, Y., Lugardon, K., Capon, C., Lopez, M., Moniatte, M., Van Dorsselaer, A., Aunis, D. and Metz-Boutigue, M.H. (1996). J. Biol. Chem. **271**, 28533–28540

23   Gadroy, P., Stridsberg, M., Capon, C., Michalski, J.C., Strub, J.M., Van Dorsselaer, A., Aunis, D. and Metz-Boutigue, M.H. (1998) J. Biol. Chem. **273**, 34087–34097

24   Gorr, S.U. and Cohn, D.V. (1990) J. Biol. Chem. **265**, 3012–3016

25   Steiner, D.F., Smeekens, S.P., Ohagi, S. and Chan, S.J. (1992) J. Biol. Chem. **267**, 23435–23438

26   Creemers, J.W., Jackson, R.S. and Hutton, J.C. (1998) Semin. Cell. Dev. Biol. **9**, 3–10

27   Portela-Gomes, G.M. and Stridsberg, M. (2001) J. Histochem. Cytochem. **49**, 483–490

28   Deftos, L.J. (1991) Endocr. Rev. **12**, 181–187

29   Metz-Boutigue, M.-H., Garcia-Sablone, P., Hogue-Angeletti, R. and Aunis, D. (1993) Eur. J. Biochem. **217**, 247–257

30   Weiler, R., Steiner, H., Fischer Colbrie, R., Schmid, K.W. and Winkler, H. (1991) Histochemistry **96**, 395–399

31   Hsiao, R.J., Mezger, M.S. and O'Connor, D.T. (1990) Kidney Int. **37**, 955–964

32   Stridsberg, M., Hellman, U., Wilander, E., Lundqvist, G., Hellsing, K. and Öberg, K. (1993) J. Endocrinol. **139**, 329–337

33   O'Connor, D.T. and Deftos, L.J. (1986) N. Engl. J. Med. **314**, 1145–1151

34   Stridsberg, M., Öberg, K., Li, Q., Engström, U. and Lundqvist, G. (1995) J. Endocrinol. **144**, 49–59

35   Wiedenmann, B. and Franke, W. (1985) Cell **41**, 1017–1028

36   Jahn, R., Schiebler, W., Ouimet, C. and Greengard, P. (1985) Proc. Natl. Acad. Sci. U.S.A. **82**, 4137–4141

37   Navone, F., Jahn, R., Di Gioia, G., Stukenbrok, H., Greengard, P. and De Camilli, P. (1986) J. Biol. Chem. **103**, 2511–2527

38   Wiedenmann, B. and Huttner, W.B. (1989) Virchows. Arch. B **58**, 95–121

39   Ozcelik, T., Lafreniere, R.G., Archer, B.T.I., Johnston, P.A., Willard, H.F., Francke, U. and Sudhof, T.C. (1990) Am. J. Hum. Genet. **47**, 551–561

40   Leube, R.E., Kaiser, P., Seiter, A., Zimbelmann, R., Franke, W.W., Rehm, H., Knaus, P., Prior, P., Betz, H., Reinke, H. et al. (1987) EMBO J. **6**, 3261–3268

41   Sudhof, T.C., Lottspeich, F., Greengard, P., Mehl, E. and Jahn, R. (1987) Science **238**, 1142–1144

42   Buckley, K.M., Floor, E. and Kelly, R.B. (1987) J. Cell Biol. **105**, 2447–2456

43   Rehm, H., Wiedenmann, B. and Betz, H. (1986) EMBO J. **5**, 535–541

44   Rubenstein, J.L., Greengard, P. and Czernik, A.J. (1993) Synapse **13**, 161–172

45   Thomas, L., Hartung, K., Langosch, D., Rehm, H., Bamberg, E., Franke, W.W. and Betz, H. (1988) Science **242**, 1050–1053

46   Bahler, M., Cesura, A.M., Fischer, G., Kuhn, H., Klein, R.L. and Da, P.M. (1990) Eur. J. Biochem. **194**, 825–830

47   Shalev, A., Gerrard, J.M., Robertson, C., Greenberg, A.H. and Linial, M. (1992) J. Cell. Biochem. **49**, 59–65

48   Kravets, F.G., Lee, J., Singh, B., Trocchia, A., Pentyala, S.N. and Khan, S.A. (2000) Prostate **43**, 169–174

49   Skibinski, G., Kelly, R.W. and James, K. (1994) Fertil. Steril. **61**, 755–759

11

Prostasomal CD46 and the anti-measles-virus effect

Masaya Kitamura*[1], Toshitugu Oka*, Kiyomi Matsumiya†, Akira Tsujimura†, Akihiko Okuyama† and Tsukasa Seya‡
*Department of Urology, Osaka National Hospital, 2-1-14 Hoenzaka, Chuo-ku, Osaka 540-0006, Japan, †Department of Urology, Osaka University Graduate School of Medicine, Suita, Japan, and ‡Department of Immunology, Osaka Medical Center for Cancer and Cardiovascular Diseases, Osaka, Japan

Introduction

Human seminal plasma has been shown to possess immunosuppressive activities [1]. E-series prostaglandins were thereafter identified as the molecules responsible for this activity, which was associated with prostasomes [2]. Prostasomes are pentalaminar or multilaminar vesicles secreted by the prostate gland. Although the physiological roles of these immunosuppressive substances are unknown, they may contribute to local modulation of the immunological environment and aid successful fertilization.

Rooney et al. [3]. reported that CD59, another inhibitor of the membrane-attack complex of complement, resides on the prostasomes. In addition, a C3-step complement-regulatory protein, decay-accelerating factor (DAF; CD55), is also present on the prostasomes, although not exclusively [4,5]. Thus the prostasomes may also be responsible for complement-mediated immune responses and their regulation by complement-regulatory proteins.

CD46, also known as membrane cofactor protein (MCP), is a C3b/C4b-binding glycoprotein that possesses cofactor activity for proteolytic inactivation of C3b and C4b to protect self cells from autologous complement attack [6]. CD46 was found to serve as a measles virus (MV) receptor [7,8]. Although CD46 was first identified [9] and then cloned [10] in leucocytes or associated cell lines, we and others have found this molecule in human semen [11,12]. A spermatozoan-membrane CD46, with a molecular mass of 43000 Da, was localized on the spermatozoan inner acrosomal membranes [11,12] and may play a role in sperm–egg interactions.

We first described a 60000 Da soluble form of CD46 that was abundant in seminal plasma (seminal plasma soluble CD46; ssCD46), and this protein was also recognized by antibodies to the membrane form of CD46 [12,13]. The roles and structural and functional properties of ssCD46 are as yet unknown. In this study, we focused on the localization and characterization of CD46 in human seminal plasma.

[1]To whom correspondence should be addressed (e-mail masaya@onh.go.jp).

Materials and methods

Samples, cells and antibodies

Semen was obtained from healthy volunteers and the seminal plasma was recovered by centrifugation (1600 g for 10 min) after liquefying for 30 min at room temperature. Prostasomes were prepared from the seminal plasma by centrifugation at 200000 g for 20 h at 4°C. The soft pellet was resuspended in PBS. Thus the prostasome samples were prepared within 25 h of ejaculation.

Chinese hamster ovary cells producing human soluble CD46 were provided as described previously [14]. CD46 molecules present in the supernatants were of the serine/threonine-rich domain (STC)/long cytoplasmic tail (CYT) phenotype covering amino acids 1–279. Monoclonal antibodies against human CD46 (M177) [15] and human CD59 (5H8) [16] were produced and purified from murine ascites as described previously.

SDS/PAGE and immunoblotting

Electrophoresis was performed under non-reducing conditions by the methods of Laemmli [17] using 10% acrylamide gels, and the samples were electrophoretically transblotted on to a nitrocellulose sheet [18]. CD46 and CD59 in the samples were then detected with the respective monoclonal antibody, horseradish peroxidase-conjugated goat anti-mouse IgG (diluted 2000-fold; Bio-Rad Laboratories, Richmond, CA, U.S.A.) and a Konica Immunoblotting Kit (Konica, Tokyo, Japan) as described previously [19].

Gel filtration

Seminal plasma was dialysed against 50 mM NaCl/20 mM sodium phosphate, pH 7.2, at 4°C overnight, and then applied to a Sephadex G-100 column equilibrated with 50 mM NaCl/20 mM sodium phosphate. This buffer was also used as an eluant. The protein concentrations in each fraction were evaluated by measurement of A_{280}. The elution profile of CD59 was monitored as a prostasome marker. The apparent quantity and cofactor activity of CD46 in each fraction were also assessed by immunoblotting and fluid-phase factor I cofactor assay (see below).

Factor I cofactor activity

A cofactor assay for detecting degradation products of C3 was performed as described previously [19,20] with slight modifications. C3(MA), which is equivalent to activated C3(C3b), was labelled with methylamine-treated fluorescence [DACM-C3(MA)], and was used as a substrate; fluorescence was incorporated into the SH residue exposed after the opening of the thioester bond in the C3d portion. Briefly, after dialysis against 50 mM NaCl/20 mM sodium phosphate, pH 7.2, at 4°C overnight, 10 μg of DACM-C3(MA), 0.8 μg of factor I [21] and 100 μl of prostasome solution were incubated for 2–10 h at 37°C in 50 mM NaCl/20 mM phosphate buffer with or without 10 μg of complement factor H [22]. The samples were subjected to SDS/PAGE under reducing conditions by adding 2% 2-mercaptoethanol. Cofactor activity was evaluated by measuring the diminution of the 120000 Da α-chain and the evolution of its cleavage products, the 75000 Da α1 fragment and the 41 000 Da fluorescent fragment, putative C3dg, using a Hitachi F2000 spectrofluorimeter [20,22].

In some experiments, 1% Nonidet P-40 (NP-40) was used as a solubilizer, and cofactor activity of relevant CD46, either the soluble or membrane form, was determined as described above. The effect of the solubilizer on cofactor activities was compared in the soluble and membrane forms.

Immunoelectron microscopy

The immunoelectron-microscopy study was conducted using a modification of the method of Watts et al. [23]. Seminal plasma was placed on Formvar-coated nickel grids for 1 min at room temperature, which were pretreated with PBS containing 2% normal goat serum, 0.5% BSA and 0.1% gelatin. After two washes with PBS the grids were incubated with the first antibody (20 μg/ml) for 1 h at 37°C. The grids were then washed again and treated for 1 h at 37°C with a goat anti-mouse IgG [F(ab′)$_2$; 1:40 dilution] secondary antibody conjugated with 10 nm gold particles (Bio Cell RL). The grids were then washed for a final time with PBS. Mouse ascites fluid with a similar protein concentration was used as a negative control. The grids, thus prepared, were counterstained with 3% uranyl acetate in 50% ethanol for 30 s and observed under a JEOL100 CX electron microscope operated at 80 kV.

Inhibition of the infective activity of MV

Vero cells were cultured in Dulbecco's modified Eagle's medium supplemented with 5% fetal calf serum and antibiotics. The Nagahata strain of MV was a gift from Dr S. Ueda (Osaka University, Osaka, Japan). MV ($10-10^4$ pfu) was treated with a soluble form of CD46 (γMCP1), intact seminal plasma, seminal plasma containing no prostasomes, or prostasomes (50 μl of a 25% solution). At 3 h intervals, a monolayer of Vero cells was incubated with the pretreated MV at $1-10^3$ pfu/well in a 24-well plate. After cells had been cultured for 3–4 days, the cytopathic effect was evaluated under a Nikon inverted microscope. The experiments were performed three times in duplicate.

Results

Association of ssCD46 with the prostasomes

Three methods were utilized to examine the relationship between ssCD46 and prostasomes. First, the prostasomes were separated from spermatozoa and seminal plasma by centrifugation and analysed by Western blotting using anti-CD46 antibody (M177; Figure 1). As we have described elsewhere [12], a single protein of 60 000 Da was recognized in the prostasome fraction of seminal plasma, which was bigger than spermatozoan-membrane CD46 of 43 000 Da in solubilized spermatozoa, and no positive band for CD46 was found after removing prostasomes in seminal plasma. Anti-CD59 antibody (5H8) was used as a monitoring marker for the prostasomes, and we confirmed the presence of ssCD46 together with CD59 using the same blotting sheet (results not shown).

Secondly, the prostasomes were separated on a molecular sieve column, and the elution profiles of CD59 (prostasome marker) and ssCD46 were

Figure 1

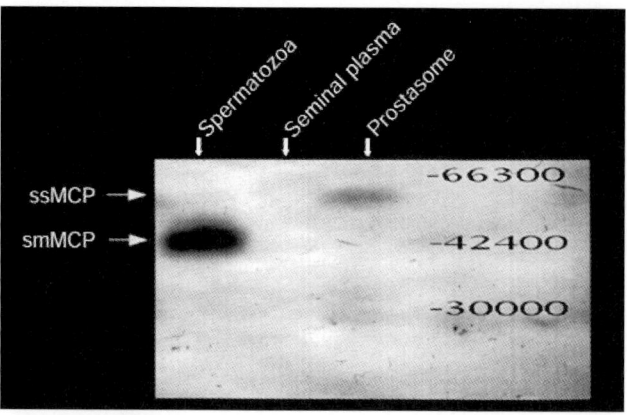

ssMCP →
smMCP →

-66300
-42400
-30000

Western blot of CD46 in semen

Solubilized spermatozoa (left-hand lane), the supernatant without prostasomes (centre lane) and the precipitate with prostasomes (right-hand lane) of seminal plasma after ultracentrifugation were electrophoresed and blotted on to a nitrocellulose sheet. The sheet was stained using anti-CD46 antibody (M177) and goat anti-mouse IgG F(ab')₂ using a Konica immunostaining kit. ssMCP, ssCD46; smMCP, spermatozoan-membrane CD46.

compared by immunoblotting (Figure 2). The prostasomes were eluted around the void volume, as shown in the blot stained with anti-CD59. ssCD46 eluted in parallel with CD59. There was a contaminating protein of 75000 Da that reacted non-specifically with the horseradish peroxidase-labelled second antibody, and ssCD46 (60000 Da) showed a faster retention time than this 75000 Da protein. The ssCD46 peak also preceded the albumin (66000 Da) peak. These findings support the idea that ssCD46 forms a complex with the prostasomes.

To confirm this further, we performed immunoelectron-microscopic analysis. M177 was used as a primary antibody for detection of CD46. CD46 molecules were identified on prostasomal membranes, as indicated by the gold particles at the periphery of the vesicles. Anti-CD59 was used as a positive control. The CD46 molecule appeared on the prostasomes, as in the case of CD59, since gold particles were seen on the prostasomal membranes in both cases (Figure 3).

Finally, we could not detect any CD46 in the seminal plasma from which the prostasomes had been removed, using a quantitative ELISA (capable of detecting 5 ng/ml CD46). Hence, the concentration of prostasome-unbound forms of CD46 is minimal (less than 5 ng/ml), if any are present at all.

Functional properties of the prostasome-bound ssCD46

A factor I cofactor assay was performed in the fluid phase using the C3d portion-labelled substrate, DACM-C3(MA). The prostasome fractions had enough cofactor activity to convert C3(MA) to C3(MA)i [the inactivated form of C3(MA)]. Since longer incubation and the addition of factor H [for preparation of C3(MA)i] revealed the possibility of further cleavage of DACM-C3(MA), the cleavage profiles of C3(MA) and C3(MA)i were chased, as shown in Figure 4. Based on the molecular mass, the smallest fluorescent fragment of C3(MA)

Figure 2

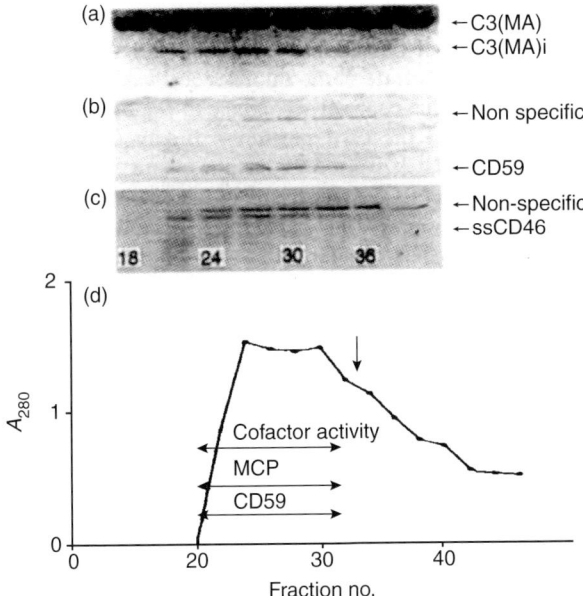

Elution profile of CD46 in seminal plasma from a Sephadex G-100 column

Cofactor activity (a), and antigens reacted with anti-CD59 antibody (5H8; b) and anti-CD46 antibody (M177; c) are shown. The latter two were the result of immunoblotting, while the former was a fluorogram using DACM-C3(MA). The positive-fraction ranges of these activities or antigens are shown in (d). The protein elution profile was read at A_{280}. The vertical arrow indicates the elution peak of BSA (66 000 Da). Cofactor activity and antigenicity of CD46 and CD59 were found in the same fractions.

corresponded to C3dg (which has a molecular mass of 41 000 Da). As with the soluble form, addition of NP-40 did not increase the cofactor activity of ssCD46, suggesting that ssCD46 resides outside of prostasomes (consistent with the electron-microscopic data) and behaves functionally as a soluble form (Figure 5).

A recombinant soluble form of CD46 (γMCP1; 30 μg) and prostasomes containing ssCD46 (50 μl of a 25% solution) were incubated with MV (10–10⁴ pfu). The infectious doses of the treated MV solutions were determined using Vero cells. A representative experiment is shown in Table 1, in which numbers of the syncytia formed were evaluated under a microscope. No blocking activity of MV infection was observed with 30 μg of γMCP1. In contrast, strong inhibition of MV infection was accomplished by the prostasomes, but little inhibitory effect was observed in other fractions of seminal plasma (Table 1).

Discussion

In this study, we have demonstrated that CD46 in human seminal plasma is bound to the prostasome, according to the results of ultracentrifugation, gel filtration and electron-microscopic analyses, and by its antigeniety and cofactor activity. The

Figure 3

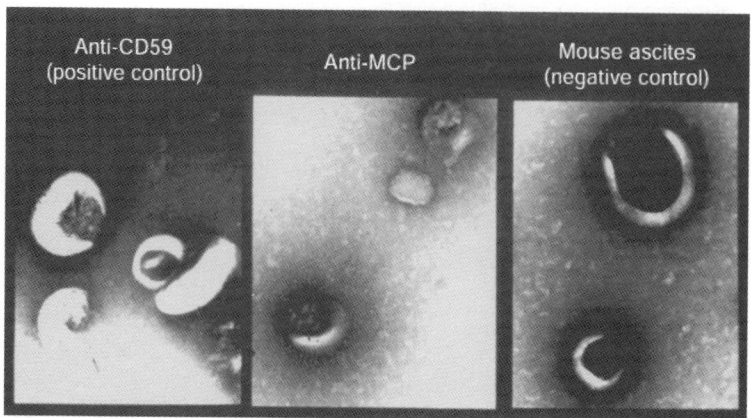

Immunoelectron-microscopic features of prostasomes stained with anti-CD59 and anti-CD46

The prostasomes were reacted with anti-CD59 antibody (5H8; left-hand panel), anti-CD46 antibody (M177; centre panel) or mouse ascites fluid (right-hand panel), and then a secondary antibody bearing gold particles (10 nm in diameter). The samples were observed under an electron microscope. Arrows indicate gold particles conjugated with goat anti-mouse IgG, which reacted with 5H8 and M177. CD46 (MCP) was found on the prostasomes, as was CD59.

results have allowed us to conclude that ssCD46 is not a proteolytic product derived from membrane forms. The soluble CD46 in human seminal plasma, now shown to be prostasomal CD46, expressed factor I cofactor activity comparable with that in recombinant soluble CD46 in the absence of NP-40. The result suggested that the molecules were expressed mostly on the prostasome surface, to function in seminal fluid. This is consistent with the report of 5′-nucleotidase on prostasomes [24]. Taken together with reports on prostasome-bound DAF and CD59 [3–5], this suggests that the prostasomes show a potential complement-regulatory profile by expressing CD46, DAF and CD59.

The roles and functions of the prostasome are unclear. Rooney et al. [3] suggested that prostasomes selectively carry glycosylphosphatidylinositol-linked proteins such as DAF and CD59 and that they have the capacity to transfer these proteins to spermatozoa [3,25]. Although some membrane-bound protein was shown to be transferred from prostasomes [26], CD46 is detected exclusively on the inner acrosomal membrane [11,12], and prostasomes cannot be its only carrier or reservoir. According to other reports, the prostasomes have membrane-linked ecto-enzyme systems, such as Mg^{2+}- and Ca^{2+}-dependent ATPase, protein kinase and zinc-dependent peptidase activities [27,28]. These enzymes may modulate the signalling pathways and confer the forward mobility of the spermatozoa [29]. In addition, fusion of prostasomes to spermatozoa might affect membrane fluidity [30] and might prevent the acrosome reaction of spermatozoa [31].

In addition to a potential complement-regulatory profile, unlike other complement receptors on prostasomes, the production of C3dg by prostasomes might indicate another function, because C3dg is a physiologically active substance,

(a)

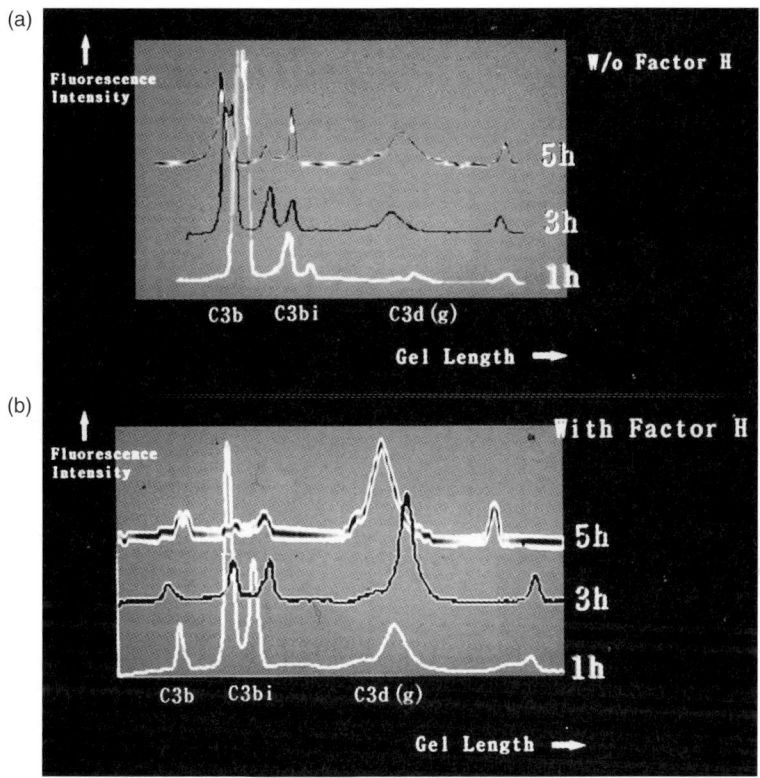

Figure 4

(b)

Evaluation of the products generated by factor I and the prostasomes

A factor I cofactor assay was performed using the substrates DACM-C3(MA) or C3(MA)i (prepared by the addition of factor H). The fragments carrying fluorescence (i.e. the C3d portion) were analysed by SDS/PAGE (under reducing conditions) using a spectrofluorimeter. (**a**) Fluorescence scanning of the electrophoretogram. The substrate DACM-C3(MA) (without factor H) was incubated with protease factor I and the prostasomal fraction containing CD46 (see Figure 2). At timed intervals, the reaction was stopped and the samples were resolved by SDS/PAGE. (**b**) The substrate DACM C3(MA)i (with factor H) was incubated with the same amounts of factor I and the prostasomal fraction as in (a). The fluorescent peaks corresponding to the C3 derivatives are indicated at the bottom of each panel, based on the mobility of each band.

increasing vascular permeability [22], inducing leucocyte chemotactic activity and suppressing T-cell proliferation [22,32]. If we regard ssCD46 as being responsible for C3dg generation, we could demonstrate a novel cofactor activity of CD46. However, naturally, the prostasomes contain various proteases [28] that might be responsible for the generation of C3dg.

A new aspect of prostasome function shown in this report is protection against microbial infection. Seminal plasma has a very strong immunosuppressive effect [1,2] because spermatozoa are alloantigenic and might be attacked by antibodies in the female genital tract. The complement system is suitable for the protection of the reproductive system because it recognizes only species, not individuals. However, many micro-organisms have adapted to the system or have

Figure 5

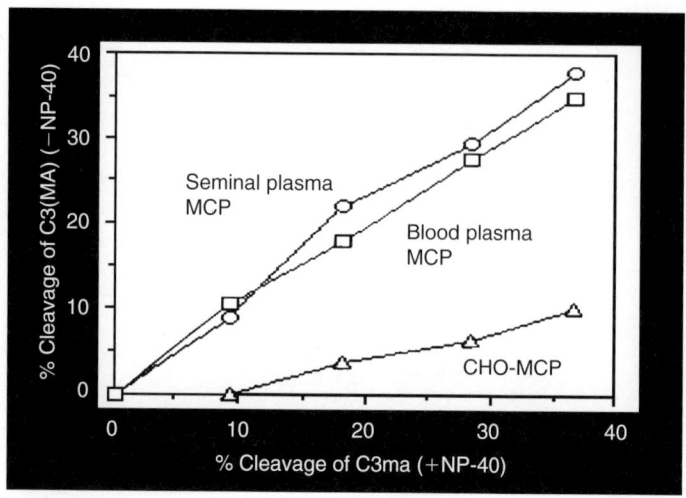

Influence of detergent (NP-40) on the cofactor activity of CD46 from various sources

Variable amounts of the membrane and soluble forms of CD46 were incubated with constant amounts of factor I and DACM-C3(MA) in buffer containing 0.5% NP-40 (x axis) or buffer without NP-40 (y axis). △, Membrane CD46(MCP) isolated from Chinese hamster ovary (CHO) cells; ○, soluble form isolated from the prostasomes in seminal plasma; □, recombinant soluble form, γMCPI.

even used it for the invasion of host cells [33]. CD46, CD55 and complement receptor-1 are specific ligands for MV [7,8], echo virus [34] and Epstein–Barr virus [35], respectively. Additionally, some complement receptors have been found to serve as bacterial receptors [36,37]. Some mycobacteria, *Listeria* spp., trypanosomes and *Leishmania* spp. deposit C3 fragments and invade host cells via indirect interactions with complement receptor proteins on the surface [38]. We first demonstrated that the prostasomes could inhibit viral activity strongly in the Vero cell system, probably via CD46, whereas small amounts of recombinant soluble form could not. A possible interpretation for these data is that prostasomes function like a 'mock cell' and, by taking up the virus, renders it unable to infect another cell. In contrast, soluble forms of CD46 bind to the MV H protein but are

Table 1

Treatment	MV (pfu)...	10^4	10^3	10^2	10
Prostasomes		0/1	0/0	0/0	0/0
Seminal plasma without prostasomes		8/4	2/1	0/0	0/0
γMCPI (soluble form)		12/9	3/3	0/0	0/0
PBS		5/3	1/0	0/0	0/0

Inhibition of syncitium formation of measles virus by prostasomes

Data indicate syncitial formation per well (in duplicate). Prostasomes, seminal plasma from which the prostasomes had been removed by centrifugation, recombinant soluble CD46 or buffer only were incubated with 10–10⁴ pfu of MV, then the mixtures were diluted 10-fold. The serially diluted MV sources were poured over the Vero cell monolayers in 24-well plates in duplicate. Then, 3 days later, the number of syncytia formed in the Vero cells was counted under a microscope. The experiments were performed three times, and a representative one is shown.

insufficient for blocking viral infection due to the lack of membrane architecture. In fact, some might argue that the prostasome, with multiple complement receptors, facilitates a 'safe' microenvironment for spermatozoa by acting as a 'trap' for viruses or other micro-organisms, which facilitate complement receptors for invasion of the host cells without affecting alloantigenic spermatozoa.

In summary, the soluble form of CD46 in semen is a prostasome-bound form which functions as the membrane CD46 in complement inactivation and probably MV adsorption. Additionally, it expresses complement-regulatory activity, even in the absence of detergents, which is enough to inhibit C3b in the fluid phase. Strong inhibition of MV infection suggests a similar function against invasion by other micro-organisms, allowing safe local 'immunity' for spermatozoa.

References

1 Stites, D.P. and Erickson, R.P. (1975) Nature (London) **253**, 727–729
2 Kelly, R.W., Holland, P., Skibinski, G. et al. (1991) Clin. Exp. Immunol. **86**, 550–556
3 Rooney, I.A., Atkinson, J.P., Krul, E.S. et al. (1993) J. Exp. Med. **177**, 1409–1420
4 Rooney, I.A. and Atkinson, J.P. (1993) Mol. Immunol. **30** (suppl. 1), 47
5 Hara, T., Matsumoto, M., Fukumori, Y. et al. (1993) Immunol. Lett. **37**, 145–152
6 Seya, T., Turner, J. and Atkinson, J.P. (1986) J. Exp. Med. **163**, 837–855
7 Dorig, R.E., Marcil, A., Chopra, A. and Richardson, C.D. (1993) Cell **75**, 295–305
8 Naniche, D., Varior-Krishnan, G., Cervoni, F. et al. (1993) J. Virol. **67**, 6025–6032
9 Cole, J.L., Housley, G.A., Dykman, T.R., Macdermott, R.P. and Atkinson, J.P. (1985) Proc. Natl. Acad. Sci. U.S.A. **82**, 859–863
10 Lublin, D.M., Liszewski, M.K., Post, T.W. et al. (1988) J. Exp. Med. **168**, 181–194
11 Anderson, D.J., Michaelson, J.A. and Johnson, P.M. (1989) Biol. Reprod. **41**, 285–293
12 Seya, T., Hara, T., Matsumoto, M. et al. (1993) Eur. J. Immunol. **23**, 1322–1327
13 Hara, T., Kuriyama, S., Kiyohara, H., Nagase, Y., Matsumoto, M. and Seya, T. (1992) Clin. Exp. Immunol. **89**, 490–494
14 Kojima, A., Iwata, K., Seya, T. et al. (1993) J. Immunol. **151**, 1519–1527
15 Seya, T., Hara, T., Matsumoto, M. and Akedo, H. (1990) J. Immunol. **145**, 238–245
16 Sugita, Y., Ito, K., Shiozuka, K. et al. (1994) Immunology **82**, 34–41
17 Laemmli, U.K. (1970) Nature (London) **227**, 680–685
18 Towbin, H., Staehelin, T. and Gordon, J. (1979) Proc. Natl. Acad. Sci. U.S.A. **76**, 4350–4354
19 Matsumoto, M., Seya, T. and Nagasawa, S. (1992) Biochem. J. **281**, 493–499
20 Seya, T., Okada, M., Nishino, H. and Atkinson, J.P. (1990) J. Biochem. (Tokyo) **107**, 310–315
21 Nagasawa, S., Ichihara, C. and Strold, R.M. (1980) J. Immunol. **125**, 578–582
22 Seya, T. and Nagasawa, S. (1985) J. Biochem. (Tokyo) **97**, 373–382
23 Watts., M.J., Dankert, J.R. and Morgan, B.P. (1990) Biochem. J. **265**, 471–477
24 Fabiani, R. and Ronquist, G. (1993) Clin. Chim. Acta **216**, 175–182
25 Rooney, I.A., Heuser, J.E. and Atkinson, J.P. (1993) Mol. Immunol. **30** (suppl. 1), 48
26 Arienti, G., Carlini, E., Verdacchi, R. et al. (1997) Biochim. Biophys. Acta **1336**, 269–274
27 Laurell, C.B., Weiber, H., Ohlsson, K. et al. (1982) Clin. Chim. Acta **126**, 161–170
28 Ronquist, G. and Brody, I. (1985) Biochim. Biophys. Acta **822**, 203–218
29 Fabiani, R., Johansson, L., Lundkvist, O. et al. (1994) Hum. Reprod. **9**, 1485–1489
30 Carlini, E., Palmerini, C.A., Cosmi, E.V. et al. (1997) Arch. Biochem. Biophys. **343**, 6–12
31 Cross, N.L. and Mahasreshti, P. (1997) Arch. Androl. **39**, 39–44
32 Meuth, J.L., Morgan, E.L., Disipio, R.G. and Hugli, T.E. (1983) J. Immunol. **130**, 2605–2611
33 Seya, T. (1995) Microbiol. Immunol. **39**, 295–305
34 Bergelson, J.M., Chan, M., Solomon, K.R. et al. (1994) Proc. Natl. Acad. Sci. U.S.A. **91**, 6245–6249
35 Fingeroth, J.D., Weis, J.J., Tedder, T.F. et al. (1984) Proc. Natl. Acad. Sci. U.S.A. **81**, 4510–4514
36 Atkinson, J.P., Krych, M., Nickells, M. et al. (1994) Clin. Exp. Immunol. **97** (suppl. 2), 1–3
37 Nowicki, B., Hart, A., Coyne, K.E. et al. (1993) J. Exp. Med. **178**, 2115–2121
38 Cooper, N.R. (1991) Immunol. Today **12**, 327–331

12

Prostasomal CD26 in fusion with sperm cells and in HIV infection

**Rafael Franco*[1], Cinzia Allegrucci†, Agustín Valenzuela‡, Julià Blanco§,
Ara G. Hovanessian‡, Francisco Ciruela*, Carmen Lluis*, Gunnar Ronquist¶
and Alba Minelli||**

*Department Bioquímica i Biologia Molecular, Facultat de Química, Martí i Franquès 1,
08028 Barcelona, Spain, †Division of Animal Physiology, School of Biosciences,
University of Nottingham, Sutton Bonington Campus, Loughborough LE12 5RD, U.K.,
‡Unité de Virologie et d'Immunologie Cellulaire, UA CNRS 1157, Institut Pasteur,
28 rue du Dr. Roux, 75724 Paris Cedex 15, France, §Fundació IRSI-Caixa, Hospital
Universitari German Trias i Pujol, Badalona, Barcelona, Spain, and ¶Department of
Medical Sciences, Clinical Chemistry, University Hospital, SE-751 85 Uppsala, Sweden,
||Dipartimento di Scienze Biochimiche e Biotecnologie Molecolari, Sezione di
Biochimica Cellulare, Via del Giochetto, 06123 Perugia, Italy,

Introduction

HIV particles block the binding of adenosine deaminase (ADA) to CD26, which is also known as dipeptidyl peptidase IV (DPP IV). The block is due to envelope glycoprotein gp120, which can be released from viruses and, therefore, can be found in various concentrations in seminal fluid. CD26/DPP IV and ADA are found in seminal vesicles. After fusion between these vesicles and horse sperm cells, an increase in ADA and CD26 levels is observed in the post-acrosomal region and at the midpiece of the spermatozoa, with a high degree of co-localization. This indicates the capturing of these molecules by sperm cells. On the other hand, pretreatment of sperm cells with exogenous ADA blocks the interaction between sperm cells and the ADA expressed in the vesicles, thus leading to a decrease in fusion. These results agree with a model of membrane–membrane recognition in which ADA could establish a bridge between vesicles and spermatozoa. Despite the controversy about the possibility that sperm cells are infected by HIV and can contribute to the spread of infection, HIV-1 viral particles and/or gp120 can affect the fusion process by blocking the ADA–CD26 interaction. In lymphocytes the ADA–CD26 interaction is very important for signal transduction and lymphocyte activation. Thus HIV, by affecting the function of the ADA–CD26 module in seminal vesicles and sperm cells, can eventually affect other processes related to fertility.

[1]To whom correspondence should be addressed (e-mail r.franco@bq.ub.es).

Prostasomes, spermatozoa and CD26/DPP IV

Prostasomes, or vesiculosomes, present at their surface a highly specific serine-type protease [1–3], which cleaves N-terminal dipeptides from peptides with a proline or alanine at the penultimate position. This enzyme is known as DPP IV (and also as CD26), first identified as a marker for lymphocyte activation. It has been hypothesized that prostasomes, by interacting closely with spermatozoa, can modify the sperm micro-environment and assist its fertilizing potential [4–7]. Sperm cells acquire glycosylphosphatidylinositol-anchored proteins present at the prostasomal surface: CD59, CD55, CDw52 or CD46. This acquisition is due partly to adherence of prostasomes to cells and partly to a second mechanism that may involve micellar intermediates [8–12].

We have reported the occurrence of a fusion-like process between prostasomes and horse spermatozoa [2]. This event, which also occurs with human prostasomes [13–15], leads to enrichment of spermatozoa in CD26/DPP IV. The event, starting with the formation of a clearly visible bridge between the two membranes, proceeds gradually towards the total embedding of the vesicle in the sperm cell membrane. This fusion, which seems to be pH-dependent, requires at least one protein at the sperm's surface, but proteins at the prostasomal surface also appear to be important [15].

CD26/DPP IV and ADA

ADA (also adenosine aminohydrolase; EC 3.5.4.4), an enzyme involved in purine metabolism, catalyses the hydrolytic deamination of adenosine or 2′-deoxyadenosine to inosine or 2′-deoxyinosine and ammonia. Although this enzyme is present in all mammalian tissues, it appears to have a major role in the development and function of lymphoid cells. Thus the level of enzyme activity in lymphocytes/thymocytes has been found to be altered in several benign and malignant diseases such as acute leukaemia [16], lymphomas [17,18], myasthenia gravis [19] and AIDS [20,21]. On the other hand, inherited deficiency of ADA leads to severe combined immunodeficiency syndrome (SCID), which is characterized by abnormal development of lymphoid tissue, lymphopenia and impairment of cell-mediated (T-cell) and humoral (B-cell) immunity [22].

In lymphocytes, as in a variety of cell types, the enzyme is located mainly in the cytosol. However, by fluorescence labelling we have found that ADA is present on the surface of a high proportion of B-lymphocytes and macrophages, and also in some T-lymphocytes from peripheral blood [23]. Previously, ADA was shown to associate with an integral membrane protein referred to as the ADA-binding protein (ADAbp), present on the cell surface of human skin fibroblasts [24]. Later, Kameoka et al. [25] showed that cell-surface ADA (ecto-ADA) binds to the T-cell activation antigen CD26, which is identical to DPP IV, a serine protease present as an ecto-enzyme in a variety of mammalian cells. There is thus an identity between ADAbp and CD26/DPP IV. Although expression of CD26 is increased in activated T-cells, its precise function in the activation has not been elucidated fully. The use of specific antibodies against different epitopes in CD26 has suggested a co-stimulatory role for CD26 in cells triggered via the T-cell receptor (for references, see [26]), and this event appears to be associated with increased activity of CD4-associated p56[lck] tyrosine kinase and enhanced phosphorylation of the ζ chain of the T-cell receptor–CD3 complex [27,28]. One of the functions of ecto-ADA might be the protection of cells against the inhibitory effect of adenosine in the culture medium [29]. Furthermore, we have shown that the interaction of ADA and CD26 on the cell surface provides a co-stimulatory signal in the activation events mediated by the T-cell receptor–CD3 complex [30]. Therefore, the multifunctional roles of ecto-ADA and CD26, as enzymes and as co-stimulatory proteins, may be essential for proper functioning of the immune system.

In lymphoid and non-lymphoid cells we have shown that there exist other cell-surface receptors for ecto-ADA [31,32]. Thus in a smooth-muscle cell line, as in other non-lymphoid cells, the A_1 adenosine receptor acts as a receptor for ecto-ADA. Therefore the enzyme can appear on the surface of cells that do not express CD26/DPP IV on the plasma membrane. To our knowledge this has been the first example of an interaction between a membrane receptor and the enzyme that degrades the endogenous ligand. Interestingly ADA not only degrades extracellular adenosine but is also needed for a high-affinity binding to the A_1 adenosine receptor. We have devised a model that can explain the physiological meaning of the interaction. At low extracellular adenosine concentrations the enzyme favours the high-affinity binding of adenosine to A_1 receptors. In fact, in the absence of the enzyme the binding of adenosine to the receptor is of low affinity. At higher concentrations, the interaction between the enzyme and the receptor is lost and, therefore, the binding of adenosine to the receptor is of low affinity and then the enzyme, whose K_m value is around 20 μM, efficiently degrades adenosine. This can be considered as a mechanism of short-term desensitization [33].

CD26/DPP IV and ADA in the fusion between seminal membrane vesicles and spermatozoa

Horse seminal vesicles and sperm cells display ecto-ADA and CD26/DPP IV. This has been demonstrated by immunostaining, immunoblotting and measurement of the enzyme activity. On the other hand, the presence of A_1 adenosine receptors in horse sperm cells but not in vesicles has been demonstrated by binding assays and by immunoblotting [34].

When fluorescent seminal vesicles are incubated with horse sperm, the fluorescence is found to be incorporated on to the plasma membrane of sperm cells. After 30 min of incubation the process is complete and the fluorescence is located mainly at the midpiece and in the post-acrosomal region [34]. Sperm cells stained after fusion with fluorochrome-conjugated antibodies showed CD26 and ADA at the midpiece and in the post-acrosomal region. A high level of co-localization is seen in Figure 1, indicating the capture of these molecules from seminal vesicles. In fact, the distribution and the intensity of the labelling for CD26 and ADA on the sperm plasma membrane is different before (moderate intensity without co-localization of ADA and CD26) and after (high intensity and

Figure 1

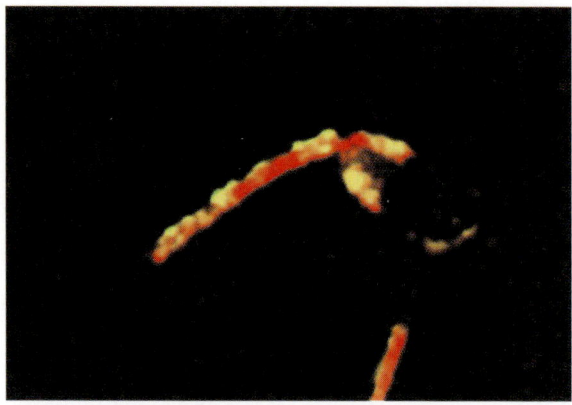

Co-localization of ADA and CD26 at the midpiece and in the post-acrosomal region of horse sperm cells after fusion with membrane vesicles

Sperm cells were incubated with vesicles at 37 °C and pH 7.4 for 30 min. Cells were adhered to glass coverslips, rinsed in PBS, fixed and permeabilized with methanol at −20 °C for 3 min. Double immunofluorescence staining using fluorescence-conjugated anti-CD26 monoclonal antibody Ta1 (Ta1-FITC; in green) and tetramethylrhodamine-conjugated anti-ADA antibody (anti-ADA-TRITC; in red) was performed as described by Minelli et al. [34]. The confocal scanning laser microscopy image also shows the co-localization of ADA and CD26 (yellow).

high degree of co-localization in the midpiece and post-acrosomal region) the fusion. In addition, the transfer of CD26 from vesicles to cells during fusion can also be followed by measuring enzyme activity. The scarce activity found in sperm cells is greatly enhanced after fusion, of the maximum activity being found after 20 min of incubation (Table 1). Moreover, pretreatment of sperm cells with exogenous ADA blocks the interaction between sperm cells and ADA expressed on seminal-vesicle membranes and, therefore, a decrease in fusion results [34].

We have devised a model of cell–cell recognition in which ADA could establish a bridge between two cells, one expressing CD26 and another expressing A_1 receptors. According to this model (see Figure 2) ADA would be involved in the recognition between sperm cells and seminal vesicles. ADA would be bound to CD26 on the membrane of seminal vesicles and would be recognized by the A_1 adenosine receptors expressed on the sperm cells. Anything able to disrupt the ADA–CD26 complexes would prevent fusion. Therefore, HIV, which disrupts ADA–CD26 interaction (see below), would affect fertilization in part by preventing seminal vesicles from fusing with the sperm cells.

Table 1

Sample	Activity (m-units/min per mg of protein)
Vesicles	6000 ± 1000
Sperm cells before fusion	6 ± 1
Sperm cells after fusion	36 ± 9

Enrichment of CD26/DPP IV activity in sperm cells after fusion with prostasome-like vesicles

Data are taken from Minelli et al. [34].

Figure 2

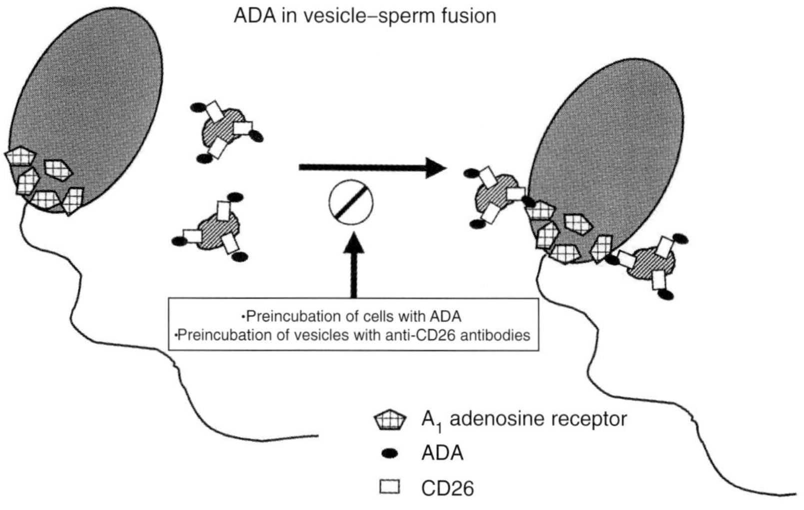

ADA in vesicle–sperm fusion

•Preincubation of cells with ADA
•Preincubation of vesicles with anti-CD26 antibodies

⬠ A$_1$ adenosine receptor
● ADA
□ CD26

ADA in vesicle–sperm fusion
Fusion can be blocked by preincubation of cells with ADA or preincubation of vesicles with anti-CD26
antibodies [34].

HIV blocks ADA–CD26 interaction in a lymphocytic cell model

Human lymphocytes are a suitable model with which to study the physiological function of ecto-ADA–CD26 complexes in relation to HIV infection. The expression of ADA in T-lymphocytes is relatively low compared with that in B-lymphocytes. In contrast, T-lymphocytes can be infected by HIV-1 particles because they carry the main receptor for the virus, CD4, whereas B-lymphocytes cannot be infected because they lack CD4. Irrespective of the possibility of infecting cells, HIV-1 particles or soluble gp120, a glycoprotein derived from the virus, can interact with a variety of membrane components other than CD4 in T- and also in B lymphocytes. Among those are the chemokine receptor CXCR4, which is considered to be a co-receptor for HIV-1 infection. CD26 is able to degrade chemokines acting on CXCR4 [35] and also interacts with CXCR4 in cells that either express CD4 or not (CD4$^+$ or CD4$^-$ cells [32]). The relevance of the effect of HIV-1 on the binding of ADA to CD26 and on the function of this complex on the surface of CD4$^+$ or CD4$^-$ cells has been studied over the last few years in our laboratory.

In experiments performed in a established human cell line and also in T-lymphocytes from peripheral blood, HIV-1 viral particles inhibit ^{125}I-labelled ADA binding to CD26 in CD4$^+$ and CD4$^-$ human cell lines [36]. Inhibition is due to the envelope glycoprotein gp120, since three different preparations of recombinant soluble gp120 were able to displace ^{125}I-labelled ADA binding to the cell surface of CD4$^+$ and CD4$^-$ cells. ^{125}I-Labelled ADA binding to these cells was displaced more than 80% of the time by unlabelled ADA, and the majority

(>90%) of specific binding was to CD26, as deduced from the decrease in the label in the presence of TA5.9, a monoclonal antibody directed against the ADA-binding site of CD26. Therefore, HIV-1 particles and soluble gp120 inhibit specific binding of ADA to cell-surface CD26 (Figure 3).

Figure 3

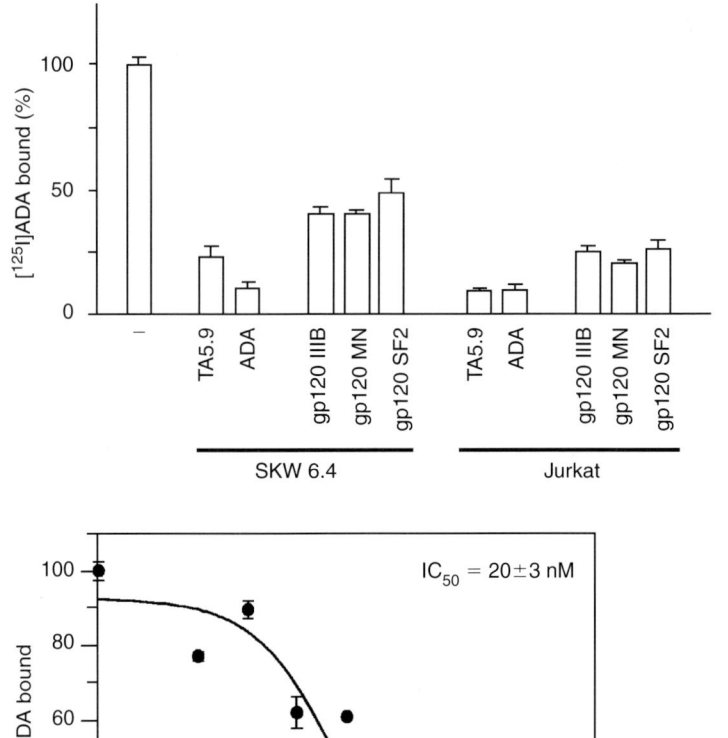

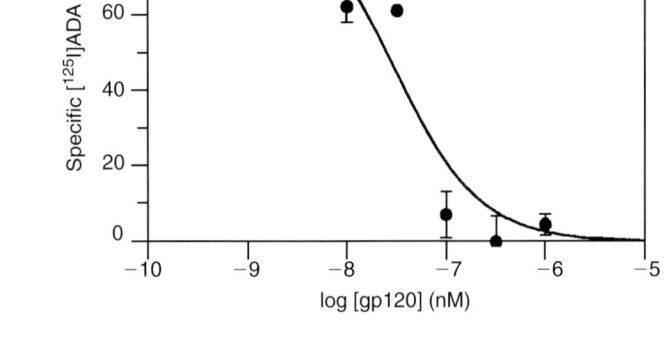

[¹²⁵I]ADA binding to CD26 is inhibited by HIV-1 envelope glycoprotein gp120

Top panel: the binding of 15 nM ¹²⁵I-labelled ADA to CD4⁺ Jurkat T-cells and to CD4⁻ SKW6.4 B cells was assayed as indicated in Valenzuela et al. [36]. Prior to addition of [¹²⁵I]ADA, cells were incubated for 15 min with medium (−) or with 1 μg/ml TA5.9 (an antibody directed against the ADA-binding site of CD26), 4 mM ADA (excess of unlabelled ADA to displace specific ¹²⁵I-labelled ADA binding) or 100 nM recombinant soluble gp120 corresponding to HIV-1 IIIB (or Lai), MN or SF2 isolates. Bottom panel: dose–response curve of gp120 IIIB inhibition of 15 nM [¹²⁵I]ADA binding to Jurkat cells.

This effect of gp120 is physiologically relevant since it occurs at subnanomolar concentrations for HIV-1 particles and at nanomolar levels for gp120. It should be noted that the concentrations of HIV-1 or gp120 in the extracellular fluids of infected individuals vary in the different stages of HIV infection. In fact, the blocking effect on the ADA–CD26 interaction can be exerted effectively by HIV-1 particles or gp120. The motif in the gp120 molecule responsible for the inhibition appears to be located in the C3 region, a constant (non-variable) region between the V3 and V4 loops. HIV-1 Lai, a T-cell-line-adapted HIV-1 isolate, was able to inhibit the binding of ADA to CD26 in a dose-dependent manner [36].

It was important to know whether this effect of gp120 is only for preventing the interaction or whether gp120 is also able to disrupt the interaction of pre-established ADA–CD26 complexes. After devising a suitable protocol, using living CD4$^+$ or CD4$^-$ cells incubated with gp120, it was possible to demonstrate that gp120 is able to displace ADA after binding itself to CD26 [32]. This HIV-induced disruption of the ADA–CD26 interaction might have functional consequences in lymphocytes and in other cell types, where the ecto-ADA–CD26 complex is present.

Physiological consequences of the blockade of ADA–CD26 interaction by HIV

Dong et al. [29] have shown that adenosine inhibits T-cell proliferation and interleukin-2 production induced by various stimuli in CD26-negative cells, which is a consequence of the lack of ecto-ADA expression. In contrast, cells expressing CD26, and thus ecto-ADA, are resistant to the inhibitory effects of adenosine. In view of these findings, it has been suggested that ADA bound to cell-surface CD26 could be responsible for a reduction in the local concentration of adenosine [29], and thus secure the homoeostasis of T-lymphocytes. Besides its role in the metabolism of adenosine, Martin et al. [30] have demonstrated that the interaction between ADA and CD26 on the cell surface of peripheral blood T-cells is essential for activation via the T-cell receptor–CD3 complex.

A further insight into the role of ecto-ADA has been obtained using cells from ADA-deficient patients. These patients have a complete lack of immune function and suffer from a SCID. T-cells from ADA-deficient SCID patients do not activate efficiently in response to antigens. Interestingly, when these cells are corrected by a gene-therapy protocol, the activation is restored and there is a high level of ecto-ADA expression. This recovery in the level of expression of ecto-ADA is even greater than that expected from the percentage of cells corrected, which is only 4–5%. This is due to the fact that intracellular ADA from corrected cells is released into the medium, where it can then interact with either corrected or non-corrected cells, since both have CD26. Once released, this ADA can interact with cell-surface CD26, thus allowing the formation of ADA–CD26 complexes even in cells that do not express the corrected gene (R. Franco et al., unpublished work). Gene therapy for SCID patients can take advantage of this fact and the level

Figure 4

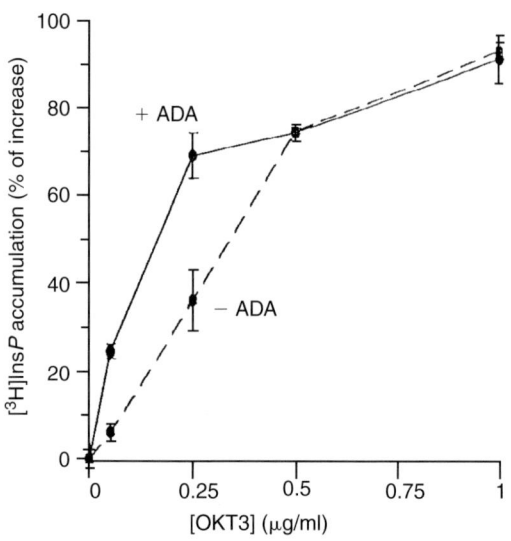

Effect of ADA on InsP accumulation in Jurkat cells triggered by activation via the T-cell receptor–CD3 complex

Cells were stimulated with the indicated amounts of the anti-CD3 antibody OKT3, in the absence (−ADA) or presence (+ADA) of 4 units/ml ADA. InsP accumulation was measured as indicated by Ciruela et al. [31].

of improvement in the functionality of peripheral blood lymphocytes can be greater than that expected from the level of expression of the corrected gene.

To confirm that this finding has a physiological relevance even in healthy individuals, experiments were performed using lymphocytes from healthy donors and a T-cell line. The activation of these normal cells was clearly modulated by exogenous ADA. ADA enhances the CD3-mediated effect on InsP levels and calcium mobilization (Figure 4). These results indicate that ADA has a physiological role as a ligand for CD26 in blood T-cells. It is suspected that the ADA–CD26 module has a similar physiological significance in other cells.

Some of the antigen-specific responses of CD4$^+$ cells are compromised early in HIV infection. To explore the possibility that blockade of the ADA–CD26 interaction by HIV-1 may have pathophysiological consequences in AIDS, we tested whether gp120 affected the effect elicited by ADA in activated T-lymphocytes. Calcium-mobilization assays in the presence of gp120 IIIB were performed in Jurkat J32 cells and in peripheral blood lymphocytes. We and others [22,23] have reported the existence of CD4-dependent and -independent sites for HIV binding to the surface of lymphocytes. Thus to dissect out the effects mediated by the gp120–CD4 interaction, cells were preincubated with Leu-3A, a monoclonal antibody that prevents gp120 from binding to CD4. In both Jurkat cells and blood lymphocytes the results were similar; gp120 was able to revert the effect caused by ADA. This reversion occurring in the presence of Leu-3A indicates that gp120 alters T-cell activation by disrupting signal transduction via

ADA–CD26. We therefore propose that the gp120-induced blockade of the ADA–CD26 interaction is one of the mechanisms causing defective T-cell function in HIV disease.

Signalling through the ADA–CD26 interaction is probably compromised in cells other than CD4$^+$ T-lymphocytes. This possibility opens up new perspectives in the search for the molecular basis of extralymphoid abnormalities found in AIDS patients. Approximately half of AIDS patients present neurological alterations whose causes are not known. Since ADA–CD26 interaction seems to be relevant for the functionality of the central nervous system [26], the involvement of gp120-induced disruption of the ADA–CD26 interaction in the genesis of these neurological problems cannot be ruled out.

New perspectives for prostasomal CD26–ADA and HIV infection

As described above, seminal vesicles display ecto-ADA and CD26 that can play an important role in the fusion of seminal vesicles with sperm cells. The presence of HIV in the seminal fluid can, through the glycoprotein gp120, alter the content of ADA in seminal vesicles and also in spermatozoa. ADA depletion in these membranes can be the cause of an alteration in fusion processes, which can lead to alterations in the sperm micro-environment and fertilization potential.

On the other hand, the results obtained with CD4$^+$ and CD4$^-$ cells indicate that ecto-ADA modulates signal transduction through its interaction with CD26. After the fusion process, there are increases in CD26 and ecto-ADA levels at the midpiece and in the post-acrosomal region. In these membranes the co-localization of both molecules is very high and we suspect that the ADA–CD26 module can modulate the activity of sperm cells. Disruption of this module by soluble gp120 or by viral particles can constitute, at least in part, a cause of decreased fertility.

This set of results is important because it provides new perspectives in the study of the alterations produced by HIV in sperm cells. It also suggests that it would be worth studying the functional role of ADA–CD26 modules in the midpiece and post-acrosomal region of sperm.

The laboratory of R.F. was supported by a joint (Echevarne Fundation and Spanish Ministry of Education) PETRI grant (PTR92/0047) administered by Fundació Bosch i Gimpera and by grants from Fondo de Investigaciones Sanitarias de la Seguridad Social (nos. 87/1389 and 91/0272) and from CIRIT-CICYT (QFN93/4423 and SAF97/0066). The laboratory of A.G.H. was supported by grants from Institut Pasteur and Agence Nationale de la Recherche sur le SIDA (ANRS). F.C. is the recipient of a grant for reincorporating scientists from the Direcció d'Universitats, de Recerca i de Societat de la Informació of the Catalan Government.

References

1   Arienti, G., Carlini, E., De Cosmo, A.M., Di Profio, P. and Palmerini, C.A. (1998) Biol. Reprod. **59**, 309–313
2   Minelli, A., Moroni, M., Martinez, E., Mezzasoma, I. and Ronquist, G. (1998) J. Reprod. Fertil. **114**, 237–243
3   de Meester, I., Vanhoof, G., Lambier, A.M. and Scharpe, S. (1996) J. Immunol. Methods **189**, 99–105
4   Ronquist, G., Nilsson, B.O. and Hejrten, S. (1990) Arch. Androl. **24**, 147–157
5   Ronquist, G. and Brody, I. (1985) Biochim. Biophys. Acta **822**, 203–218
6   Stegmayr, B. and Ronquist, G. (1982) Scand. J. Urol. Nephrol. **16**, 85–90
7   Kelly, R.W., Hollamd, P., Skibinski, G., Harrison, C., McMillan, L., Hargreave, T. and Jam, K. (1991) Clin. Exp. Immunol. **86**, 50–58
8   Rooney, I.A., Atkinson, J.P., Krul, E.S., Schonfeld, G., Polakoski, K., Saffitz, J.E. and Morgan, B.P. (1993) J. Exp. Med. **177**, 1409–1420
9   Rooney, I.A., Heuser, J.E. and Atkinson, J.P. (1996) J. Clin. Invest. **97**, 1675–1686
10   Klappe, K., Wilschut, J., Nir, S. and Hoekstra, D. (1986) Biochemistry **25**, 8252–8260
11   Arts, G.J.M., Jager, S. and Hoekstra, D. (1994) Biochem. J. **304**, 211–218
12   Arts, G.J.M., Kuiken, J., Jager, S. and Hoekstra, D. (1993) Eur. J. Biochem. **210**, 1001–1009
13   Arienti, G., Carlini, E. and Palmerini, C.A. (1997) J. Membr. Biol. **155**, 89–94
14   Arienti, G., Polci, A., Carlini, E. and Palmerini, C.A. (1997) FEBS Lett. **410**, 343–346
15   Carlini, E., Palmerini, C.A., Cosmi, E.V. and Arienti, G. (1997) Arch. Biochem. Biophys. **343**, 6–12
16   Sylwestrowicz, T.A., Ma, D.D., Murphy, P.P., Massaia, M., Prentice, H.G., Hoffbrand, A.V. and Greaves, M.F. (1982) Leuk. Res. **6**, 475–482
17   Vezzoni, P., Giardini, R., Lucchini, R., Lombardi, L., Vezzoni, M.A., Besana, C. and Clerici, L. (1985) Am. J. Hematol. **19**, 219–227
18   Murray, J.L., Perez-Soler, R., Bywaters, D. and Hersh, E.M. (1986) Am. J. Hematol. **21**, 57–66
19   Vezzoni, P., Fiacchino, F., Clerici, L., Sghirlanzoni, A., Cerrato, D., Peluchetti, D., Lucchini, R., Raineri, M. and Cornelio, F. (1984) J. Neuroimmunol. **6**, 427–433
20   Lane, H.C., Depper, J.M., Greene, W.C., Whalen, G., Waldmann, T.A. and Fauci, A.S. (1985) N. Engl. J. Med. **313**, 79–84
21   Murray, J.L., Loftin, K.C., Munn, C.G., Reuben, J.M., Mansell, P.W. and Hersh, E.M. (1985) Blood **65**, 1318–1324
22   Hershfield, M.S. and Mitchell, B.S. (1995) in The Metabolic and Molecular Bases of Inherited Disease, vol. II (Scriver, C.R., Beaudet, A.L., Sly, W.S. and Valle, D., eds), pp. 1725–1768, McGraw Hill, New York
23   Aran, J.M., Colomer, D., Matutes, E., Vives-Corrons, J.L. and Franco, R. (1991) J. Histochem. Cytochem. **39**, 1001–1008
24   Andy, R.J. and Kornfeld, R. (1984) J. Biol. Chem. **259**, 9832–9839
25   Kameoka, J., Tanaka, T., Nojima, Y., Schlossman, S.F. and Morimoto, C. (1993) Science **261**, 466–469
26   Fleischer, B. (1994) Immunol. Today **15**, 180–184
27   Torimoto, Y., Dang, N.H., Vivier, E., Tanaka, T., Schlossman, S.F. and Morimoto, C. (1991) J. Immunol. **147**, 2514–2517
28   Mittrücker, H.W., Steeg, C., Malissen, B. and Fleischer, B. (1995) Eur. J. Immunol. **25**, 295–297
29   Dong, R.-P., Kameoka, J., Hegen, M., Tanaka, T., Xu, Y., Schlossman, S.F. and Morimoto, C. (1996) J. Immunol. **156**, 1349–1355
30   Martin, M., Huguet, J., Centelles, J.J. and Franco, R. (1995) J. Immunol. **155**, 4630–4643
31   Ciruela, F., Saura, C., Canela, E.I., Mallol, J., Lluis, C. and Franco, R. (1996) FEBS Lett. **380**, 219–223
32   Herrera, C., Morimoto, C., Blanco, J., Mallol, J., Arenzana, F., Lluis, C. and Franco, R. (2001) J. Biol. Chem. **276**, 19532–19539
33   Franco, R., Ferré, S., Agnati, L., Torvinen, M., Ginés, S., Hillion, J., Casadó, V., Lledó, P.M., Zoli, M., Lluis, C. and Fuxe, K. (2000) Neuropsychopharmacology **23**, S50–S59
34   Minelli, A., Allegrucci, C., Mezassoma, I., Ronquist, G., Lluis, C. and Franco, R. (1999) Biol. Reprod. **61**, 802–808
35   De Meester, I., Korom, S., Van Damme, J. and Scharpe, S. (1999) Immunol. Today **20**, 367–375
36   Valenzuela, A., Blanco, J., Callebaut, C., Jacotot, E., Lluis, C., Hovanessian, A.G. and Franco, R. (1997) J. Immunol. **158**, 3721–3729

13

Antioxidant properties of prostasomes

Fabrice Saez
Laboratoire de Biologie de la Reproduction, Université d'Auvergne, 28, Place Henri Dunant, 63000 Clermont-Ferrand, France

Introduction

The importance of oxidative stress as a cause of male infertility has been a growing idea for the last 20 years. It is now accepted that high amounts of reactive oxygen species (ROS) in human semen can be deleterious for sperm function [1,2]. Conversely, low doses of ROS and particularly of the superoxide anion ($O_2^{-\cdot}$) are implicated in the initiation of capacitation and the acrosome reaction (AR), two prerequisites for the occurence of fertilization [3,4]. Spermatozoa have long been considered as the major ROS-producing cells in semen but it is now clear that white blood cells, which sometimes infiltrate semen, are the main producers of ROS [5–7].

Seminal plasma possesses antioxidant properties to protect spermatozoa against ROS-induced damage, but these capacities can sometimes be overwhelmed [8]. Indeed, ROS can cause lipoperoxidation of the male gamete's plasma membrane due to its high polyunsaturated fatty acid content [9], and also attack DNA and provoke its fragmentation [10].

Leucocytospermia [the incidence of $>1\times10^6$ polymorphonuclear neutrophils (PMN)·ml^{-1}] is often a consequence of the inflammation of one or more of the male accessory sex glands [7]. However, it is not always associated with a decrease of sperm function *in vivo* [11]. The important capacity of PMN to produce ROS arises from a specialized enzyme, NADPH oxidase, which catalyses the production of $O_2^{-\cdot}$ by the reduction of oxygen, using NADPH as the electron donor:

$$2O_2 + NADPH \longrightarrow 2O_2^{-\cdot} + NADP^+ + H^+ \qquad (1)$$

This enzyme exists in an inactive conformation in the resting cells, with cytoplasmic and membrane subunits, and is assembled into an active form upon activation by stimuli such as interleukins (e.g. interleukin-8), the C5a fraction of complement or formylated peptides (reviewed by Babior [12]). In semen, the infiltrating PMN seem to be, at least in part, in an activated state [8], suggesting that they probably come from an inflammation site, and thus represent a source of oxidative stress for spermatozoa.

Correspondence should be addressed to Laboratoire BDR-CECOS, CHU, Hôtel-Dieu, BP 69, Boulevard Léon Malfreyt, 63003 Clermont-Ferrand Cedex, France (e-mail fabrice.saez@u-clermont1.fr).

Prostasomes were shown previously to have the ability to decrease $O_2^{-\cdot}$ production by blood PMN stimulated by either PMA or formyl-Met-Leu-Phe (fMLP) [13]. However, these authors did not investigate the effect of prostasomes on ROS production by seminal PMN, nor the mechanism of action of prostasomes.

In this chapter, my findings concerning the action of prostasomes on ROS production by blood and seminal PMN are summarized, together with preliminary results from studies investigating the possible influence of prostasomes on ROS production by human spermatozoa. Briefly, prostasomes were shown to inhibit ROS production by seminal PMN, via the inhibition of their NADPH oxidase activity. This inhibition was associated with a decrease in the membrane fluidity of the PMN, implicating the lipid composition of prostasomes in their mechanism of action. Prostasomes also inhibited NADPH-induced $O_2^{-\cdot}$ production in spermatozoa, measured by lucigenin-enhanced chemiluminescence. Under these conditions, they also favoured sperm capacitation.

Prostasomes and PMN

Based on the work by Skibinski et al. [13], the first step in the study consisted of evaluating the influence of prostasomes on ROS production by seminal PMN. The total ROS production of cell suspensions isolated from PMN-containing semen was measured by luminol chemiluminescence [14]. The level of ROS production was related to the quantity of PMN present in the sample, and a physiological concentration of prostasomes in the incubation medium significantly inhibited ROS production in the basal state as well as after stimulation with PMA. These results were in accordance with those in the literature [5–7] showing that PMN are the major source of ROS in human semen. The fact that prostasomes inhibited basal ROS production suggests a physiological role *in vivo* in cases of leucocytospermia. The antioxidant capacity is due to an unusual mechanism, as it has been demonstrated in my laboratory that it was not a scavenging capacity. Indeed, prostasomes could not scavenge ROS produced constantly in an aqueous solution by a particular molecule, ABAP (2,2'-azobis-2-amidinopropane dihydrochloride).

The inhibition of ROS production by PMN is associated with an effect on the membrane dynamics of these cells. When it is in contact with prostasomes, the plasma membrane of blood PMN (used as a model) is rigidified, as shown using an ESR method ([14]; Figure 1). Blood PMN were used in this case because the method requires pure cell suspensions, and it was not possible to obtain such suspensions of seminal PMN, as spermatozoa were always contaminating the samples. Our hypothesis is that the rigidification could be due to lipid transfer from the prostasomes to the spermatozoa. The transferred lipids could be cholesterol and sphingomyelin mainly, which are present in high amounts in prostasomes [15].

As the main ROS-producing enzyme of PMN is NADPH oxidase, we then investigated the influence of prostasomes on its activity. Enhanced chemiluminescence was used, with MCLA [2-methyl-6-(*p*-methoxyphenyl)-3,7-dihydroimidazo[1,2-*a*]pyrazin-3-one] as a probe. This molecule is specific to $O_2^{-\cdot}$ produced in the extracellular compartment, and is also very sensitive. The $O_2^{-\cdot}$ production of blood PMN stimulated with PMA was thus evaluated ([15]; Figure 2).

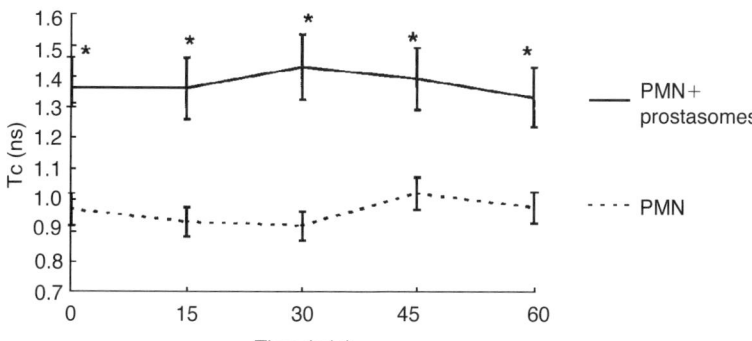

Figure 1

Time-course evolution of the correlation-relaxation time (Tc) of blood PMN plasma membrane in a medium with or without prostasomes

*Tc is inversely related to membrane fluidity and was measured using ESR with a 16-doxyl-stearic acid probe. Before addition of the prostasomes (equivalent to 130 μM cholesterol), Tc had a similar value in the two aliquots of PMN. Each value is the mean ±S.E.M. from four independent assays. *$P<0.05$ compared with the corresponding reference value. From Saez, F. Motta, C., Boucher, D. and Grizord, G. (1998) Mol. Hum. Reprod. 4, 667–672. © European Soceity of Human Reproduction and Embryology. Reproduced by permission of Oxford University Press/Human Reproduction.*

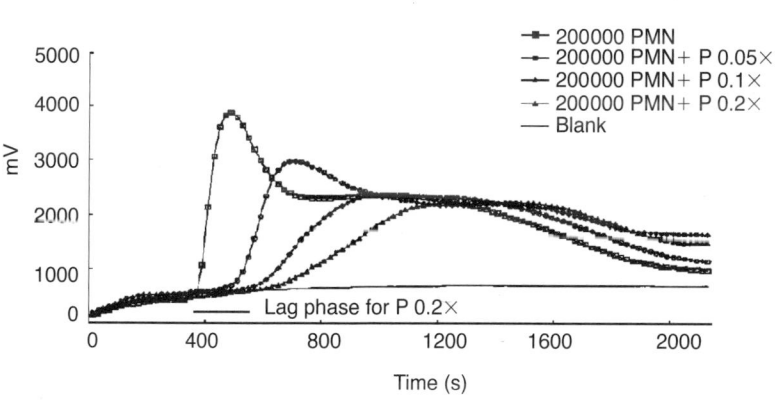

Figure 2

MCLA chemiluminescence of blood PMN stimulated with 80 nM PMA: effects of different prostasome (P) concentrations

A similar profile was obtained for luminol chemiluminescence after PMA stimulation. The curves obtained for seminal PMN had the same profile. This graph is representative of 6 and 10 different experiments for blood and seminal PMN, respectively. P 0.05×–P 0.2×, fractions of the physiological prostasome concentration (1×, equivalent to 130 μM cholesterol). From Saez, F. Motta, C., Boucher, D. and Grizord, G. (2000) Mol. Hum. Reprod. 6, 883–891. © European Soceity of Human Reproduction and Embryology. Reproduced by permission of Oxford University Press/Human Reproduction

Under these conditions, the lag phase and the total activity of the enzyme were inhibited by prostasomes. The lag phase is the characteristic activation parameter of the enzyme and corresponds to its assembly in an active conformation in secretory vesicles and to the exocytosis of these vesicles [16,17]. Normally, the lag phase of blood PMN is around 30 s. The presence of prostasomes provoked a concentration-dependent increase in the lag phase. The total enzyme activity was also decreased significantly as a function of the prostasome concentration. The inhibitory action of prostasomes was conserved if they were boiled previously for 5 min, thus suggesting that the inhibitor is not a protein.

The experiments carried out on blood PMN permitted us to validate the protocols using a well-known enzymic system. Applying the methods to seminal PMN contributed to better characterization of these cells. The NADPH oxidase activity of cell suspensions that had been isolated from leucocytospermic samples and stimulated with PMA was similar to that of blood PMN, showing that seminal PMN possess the functional enzyme. The kinetic and activation characteristics were also comparable with blood NADPH oxidase, with lag phases of 31 ± 7 s (seminal) and 32 ± 2 s (blood) [15]. On the contrary, the total enzyme activity was slightly lower for seminal PMN, which may have been due to a difference in the regulation of enzyme activity after PMA stimulation between these two cell types.

When blood PMN were stimulated with fMLP, NADPH oxidase activity was also inhibited significantly by prostasomes. In this case, however, for seminal PMN, the presence of prostasomes did not lead to significant inhibition of the enzyme activity, for which very important inter-individual variation was apparent. This variation may be attributed to variable membrane quality of the cells, in relation to their origin or previous contact with prostasomes, as we could only work with ejaculated semen.

The inhibitory effect of prostasomes on ROS production by blood PMN is the consequence of decreased NADPH oxidase activity. The inhibition was verified for two different stimuli, PMA and fMLP. For seminal PMN, there was no significant inhibition of the fMLP-stimulated NADPH oxidase activity, although there was high inter-individual variation. This result could reflect different regulation of the seminal PMN enzyme when a physiological transduction pathway is activated in semen during leucocytospermia.

Prostasomes act indirectly to protect sperm against oxidative stress. This property is in addition to the strong antioxidant capacity of seminal plasma but is of a different origin. Prostasomes act on the ROS-producing enzyme (NADPH oxidase) whereas seminal plasma contains mainly scavenging molecules or enzymes.

Prostasomes and spermatozoa

Controlled ROS production is required for capacitation and AR of spermatozoa, two events that are fundamental to their function. In contrast, excessive ROS production is toxic for sperm cells. Considering the results obtained with PMN, a role for prostasomes in the capacitation process, via the level of oxidative stress, could logically be envisioned.

We used NADPH to stimulate $O_2^{-\bullet}$ production of two sperm populations isolated by migration on Percoll gradients, to investigate a potential effect of prostasomes on this production. Spermatozoa from the 95% Percoll fraction and from the 95/65% interface fraction of a discontinuous Percoll gradient were incubated with 2.5 mM NADPH, and $O_2^{-\bullet}$ production was measured by lucigenin-enhanced chemiluminescence. Under these conditions, capacitation was estimated after a 3.5 h incubation period, using AR stimulated by the calcium ionophore A23187 (A23187-AR). In the literature, different effects have been attributed to the presence of NADPH in the incubation medium during capacitation. It was shown to trigger $O_2^{-\bullet}$ production by spermatozoa as well as capacitation of Percoll-purified sperm cells [18]. On the contrary, a negative effect of NADPH on the capacitation of sperm cells incubated under 20% O_2 was demonstrated, whereas no effect was seen under 5% O_2 [2]. Finally, NADPH was reported to stimulate capacitation without triggering any $O_2^{-\bullet}$ production [19].

Our preliminary results showed that prostasomes had the ability to decrease NADPH-induced $O_2^{-\bullet}$ production by the two sperm populations studied (Figure 3). The presence of NADPH did not have any significant influence on sperm capacitation, although this did have a tendency to decrease. When NADPH was included in the incubation medium, the presence of prostasomes led to a significant increase in capacitation compared with the same conditions without prostasomes (Figure 3). This result is specific to the conditions created by NADPH, as we verified that prostasomes inhibited the capacitation of spermatozoa under normal conditions (results not shown), in accordance with the work of Cross and Mahasreshti [20].

These results also showed a negative correlation between NADPH-induced $O_2^{-\bullet}$ production and capacitation of the sperm cells (Figure 4). Under these conditions of oxidative stress, prostasomes appear to exert protection on the spermatozoa: they inhibit their $O_2^{-\bullet}$ production and favour their capacitation.

Our results are the first showing an antioxidant function of prostasomes in relation to spermatozoa. This effect was associated with an enhancement of the final state of the capacitation process, evaluated by the AR. However, these results need to be taken with caution, because it was shown recently (after our experiments were performed) that lucigenin could not be used as a tool to measure NADPH-dependent $O_2^{-\bullet}$ production [21]. The authors showed that the enzymic system producing $O_2^{-\bullet}$ in spermatozoa is probably not an NADPH oxidase, as hypothesized previously [22]. The mechanism of action of NADPH on capacitation needs to be reviewed, as $O_2^{-\bullet}$ is probably not involved. Whatever this mechanism, we have shown a protective effect of prostasomes on capacitation when NADPH was present. Even if the measured chemiluminescence is an artifact, its intensity is related to the penetration rate of NADPH in spermatozoa, in association with their membrane quality [18]. Indeed, we obtained a higher chemiluminescence for interface spermatozoa, and this fraction presented a lower percentage of morphologically normal cells compared with the spermatozoa in the 95% Percoll fraction (results not shown). The tendency of NADPH to decrease A23187-AR could thus be due to the level of NADPH penetrating in the spermatozoa. Prostasomes in the medium might interfere with this entry, by either capturing part of the NADPH or interacting with spermatozoa, thus lowering its penetration. The mechanism of

Figure 3

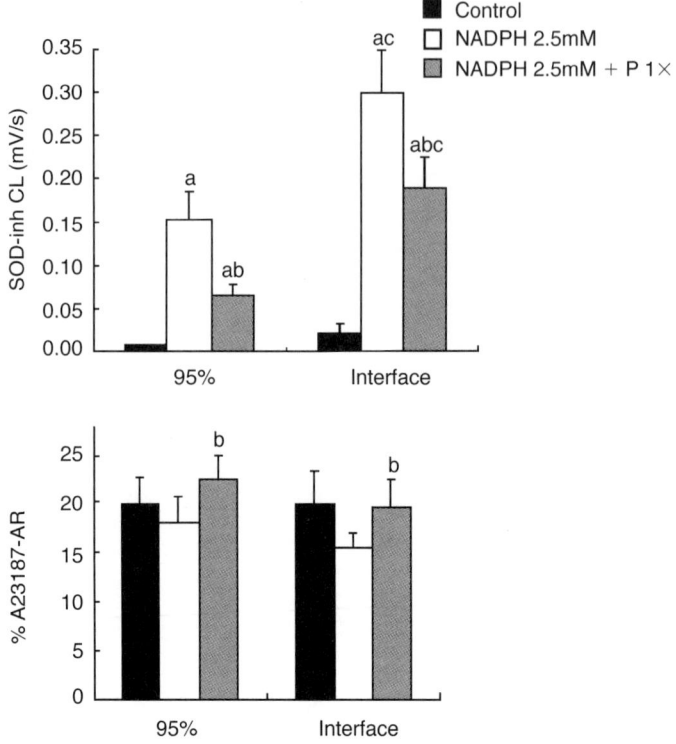

Influence of 2.5 mM NADPH and of prostasomes on the chemiluminescence inhibitable by superoxide dismutase (SOD-inh CL) and the percentage of A23187-AR of spermatozoa from the pellet (95%) and the 95/65% interface fractions of the Percoll gradients

Values are expressed as means ±S.E.M. from 11 and 9 independent samples for ROS production and A23187-AR, respectively. a, P<0.05 compared with the respective control; b, P<0.05 compared with 2.5 mM NADPH; c, P<0.05 compared with the same category in the 95% fraction; P 1×, physiological prostasome concentration (130 μM cholesterol)

action of prostasomes needs to be investigated further, but they protect the sperm's function from the deleterious effects of NADPH.

Conclusion and perspectives

Our work has confirmed the possibility that prostasomes play an antioxidant role in seminal plasma, a hypothesis raised in 1992 by Skibinski et al. [13]. Furthermore, we did not elucidate the precise mechanism of prostasome action in this process. The antioxidant function of prostasomes is potent in both leucocytes and spermatozoa. If the rule about prostasomes concerning leucocytes seems clear, i.e. limitation of oxidative stress, it is not the case for spermatozoa. Indeed, prostasomes were shown

Figure 4

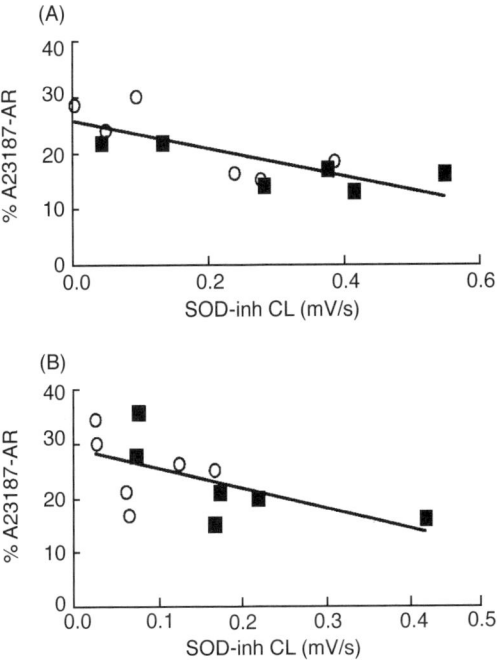

Correlation between NADPH-dependent O$_2^{-\cdot}$ production and the percentage of A23187-AR of sperm cells, in the absence (A) or presence (B) of a physiological concentration of prostasomes (P 1×)

The NADPH concentration was 2.5 mM. ○, Cells from the 95% Percoll fraction (n=6); ■, cells from the interface (n=6). In each case, the correlation is significant: r= −0.787, P=0.0024 (A) and r= −0.585, P=0.045 (B). SOD-inh CL, chemiluminescence inhibitable by superoxide dismutase.

to inhibit the capacitation process under capacitating conditions [20], but they appear, under our conditions, to protect the functional capacities of the sperm cells. This controversy needs to be investigated with further experiments.

The lipid structure and composition of prostasomes is the key determinant of their antioxidant function. We confirmed that their action on leucocytes is probably dependent on their lipids, and we assume that it is the same for their action on spermatozoa.

This work offers interesting perspectives. The mechanism of action of prostasomes on the NADPH oxidase of PMN needs to be defined precisely. Also, the role of prostasomes in the function of sperm cells, particularly in the capacitation step, needs to be investigated further. I believe that prostasomes could be a good model for the study of the capacitation process.

I thank Professor Daniel Boucher for the welcome I received in his laboratory during my Ph.D., and Dr Geneviève Grizard for being my tutor. Many thanks go to all the technical staff of the laboratory, without whom this work would not have been realized. In particular, I thank Claudine Nouailles, Christine Artonne and Monique Etienne, who helped me to perform these experiments.

References

1   Aitken, R.J. (1999) J. Reprod. Fertil. **115**, 1–7
2   Griveau, J.F. and Le Lannou, D. (1997) Int. J. Androl. **20**, 61–69
3   De Lamirande, E. and Gagnon, C. (1995) Free Radicals Biol. Med. **18**, 487–495
4   Griveau, J.F., Renard, P. and Le Lannou, D. (1995) Int. J. Androl. **18**, 67–74
5   Aitken, R.J. and West, K. (1990) Int. J. Androl. **13**, 433–451
6   Whittington, K. and Ford, W.C.L. (1999) Int. J. Androl. **22**, 229–235
7   Wolff, H. (1995) Fertil. Steril. **63**, 1143–1157
8   Aitken, R.J., Buckingham, D., Brindle, J., Gomez, E., Baker, H.W. and Irvine, D.S. (1995) Hum. Reprod. **10**, 2061–2071
9   Jones, R., Mann, T. and Shevins, R.J. (1979) Fertil. Steril. **31**, 531–537
10   Twigg, J., Irvine, D.S., Houston, P., Fulton, N., Michael, L. and Aitken, R.J. (1998) Mol. Hum. Reprod. **4**, 439–445
11   Tomlinson, M.J., Barratt, C.L.R. and Cooke, I.D. (1993) Fertil. Steril. **60**, 1069–1075
12   Babior, B.M. (1999) Blood **93**, 1464–1476
13   Skibinski, G., Kelly, R.W., Harkiss, D. and James, K. (1992) Am. J. Reprod. Immunol. **28**, 97–103
14   Saez, F., Motta, C., Boucher, D. and Grizard, G. (1998) Mol. Hum. Reprod. **4**, 667–672
15   Saez, F., Motta, C., Boucher, D. and Grizard, G. (2000) Mol. Hum. Reprod. **6**, 883–891
16   Kobayashi, T., Robinson, J.M. and Seguchi, H. (1998) J. Cell Sci. **111**, 81–91
17   Vaissiere, C., Le Cabec, V. and Maridonneau-Parini, I. (1999) J. Leukocyte Biol. **65**, 629–634
18   Aitken, R.J., Fisher, H.M., Fulton, N., Gomez, E., Knox, W., Lewis, B. and Irvine, S. (1997) Mol. Reprod. Dev. **47**, 468–482
19   De Lamirande, E., Harakat, A. and Gagnon, C. (1998) J. Androl. **19**, 215–225
20   Cross, N.L. and Mahasreshti, P. (1997) Arch. Androl. **39**, 39–44
21   Richer, S.C. and Ford, W.C.L. (2001) Mol. Hum. Reprod. **7**, 237–244
22   Aitken, R.J., Harkiss, D., Knox, W., Paterson, M. and Irvine, D.S. (1998) J. Cell Sci. **111**, 645–656

14

Interaction and fusion of prostasomes and spermatozoa

Giuseppe Arienti*[1], Enrico Carlini* and Carlo Alberto Palmerini†
*Sezione di Biochimica, Dipartimento di Medicina Interna, Università di Perugia, Via del Giochetto, 06122 Perugia, Italy, and †Dipartimento di Scienze Biochimiche e Biotecnologie Molecolari, Università di Perugia, Perugia, Italy

Introduction

Prostasomes have been described in human semen [1], but they are probably also present in the semen of other mammals [2,3]. The research on prostasomes has mainly dealt with the problem of their function in reproductive human physiology. Several possibilities have been taken into account, including sperm motility [4], immune response [5–7], semen liquefaction [8] and anti-bacterial [9] and anti-cancer [10] properties.

Prostasome composition

The composition of prostasomes has been studied in several laboratories [11,12]. Several aspects of prostasome lipids are peculiar: they show a very high cholesterol/phospholipid ratio and sphingomyelin is the main phospholipid (Figure 1). For this reason, they differ greatly from most biological membranes. The composition of fatty acids in the total phospholipid shows that prostasomes are rich in saturated species: more that 50% of prostasome phospholipid fatty acids are palmitic and stearic acids, which is in agreement with their low fluidity [12,13]. It should also be noted that phosphatidylcholine mainly possesses saturated species; the only important unsaturated fatty acid in this lipid appears to be oleic acid. Long and unsaturated fatty acids occur in sphingomyelin, that is very low in oleic acid.

The most striking point about prostasome lipids is the high amount of cholesterol. This fact is intriguing because of the relationship between cholesterol and human reproduction. Cholesterol has in fact been claimed to be a decapacitating agent [14–16].

Prostasomes also contain a number of proteins, many of them endowed with biological activity [17]. Therefore, prostasomes contain lipid and protein that may affect sperm function. The possibility of transferring this material to spermatozoa may be of some consequence for their properties.

[1]To whom correspondence should be addressed (e-mail arienti@unipg.it).

Figure 1

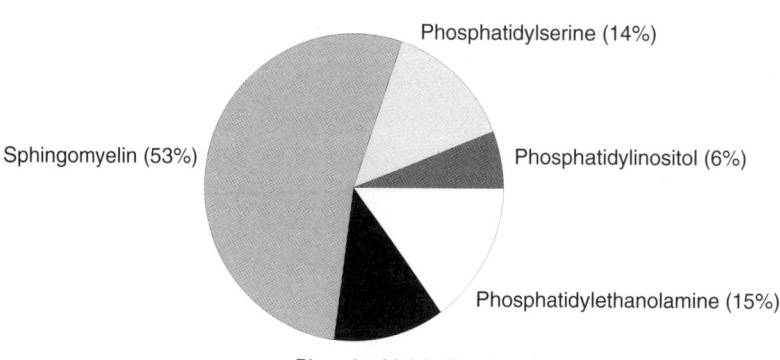

Distribution of total lipid and phospholipid phosphorus in prostasomes
Results are expressed as percentages of total lipid or total lipid phosphorus.

Transfer of lipid from prostasome to sperm

The transfer of lipid between prostasomes and sperm can be inferred from the behaviour of a fluorescent probe, octadecylrhodamine (R_{18}), which has been used widely to study membrane fusion [18]. The probe is fat-soluble and shows fluorescence self-quenching that can be relieved upon dilution. In the proper experimental setting, the fluorescent signal is inversely proportional to the concentration of probe in the lipid phase of the membrane.

Measurements of R_{18} fluorescence are performed upon mixing prostasomes, which have been labelled with the probe, with unlabelled sperm. The transfer of the probe to sperm dilutes it and a relief of fluorescence self-quenching ensues. The increase in fluorescence can be compared with the increase due to the addition of known amounts of detergent, which represents the total mixing of lipid phases. Results are therefore usually expressed as the percentage of total theoretical lipid mixing (Table 1).

pH	Fusion (%)	Table 1
5.0	25±3	
5.5	20±2	
6.0	15±3	
6.5	10±2	
7.0	6±3	
7.5	1±2	
8.0	1±2	

Fusion between prostasomes and spermatozoa after mixing at the indicated pH values for 10 min

Results are expressed as percentage of complete fusion, as described in [37].

We have shown that the dye moved from prostasomes to sperm, thus indicating the mixing of the lipid phases. The transfer was pH-dependent and could be detected at pH 7, although it was not detectable at pH 8 (Table 1). The extent of transfer increased with decreasing pH and it was routinely measured at pH 5.0. This fusion did not depend upon Ca^{2+}, only on pH value [19]. The pH-dependent transfer of lipid between sperm and prostasome relies on the integrity of both the prostasomal and spermatozoan proteins. Indeed, treatments that that are able to alter protein (protease treatment, boiling and extraction of lipid followed by liposome preparation) also affect fusion, by reducing it. To completely abolish fusion it is necessary to treat both spermatozoa and prostasomes [19].

The microscopic appearance of a mixture of prostasomes and spermatozoa at pH 5 and 8 (Figure 2) confirmed the data obtained from R_{18} self-quenching studies. Indeed, the fluorescent probe was transferred to spermatozoa only after mixing with loaded prostasomes at pH 5. The result was confirmed further by flux cytofluorimetry [19].

The pH-dependent lipid transfer between prostasomes and sperm raises some questions about the physiological role of this phenomenon. The pH values of semen (pH 7.6 [20]) are too high for fusion to occur, but the pH of ejaculated material could be lower in the vagina, at least enough to trigger fusion (pH ≈ 7). On the other hand, seminal fluid is formed from the secretions of several glands and, in the so-called split ejaculation, sperm are ejaculated first with a rather acidic prostatic secretion. This would permit contact between spermatozoa and prostasomes in more acidic conditions than those that could be inferred by measuring the pH of the whole ejaculate.

The duration of mixing may be another point to consider. The duration of contact between spermatozoa and prostasomes may be short; sperm will leave the vagina and enter the uterus, whereas prostasomes do not do this. Thus it should be considered that prostasomal–spermatozoan fusion is a fast process that is completed in a few minutes. Therefore, some material could be transferred *in vivo*. Moreover, it should be considered that the sperm plasma membrane shows a lateral patching of lipid [21,22]. Therefore, the lipid transfer between sperm and prostasomes might affect only a small proportion of total sperm plasma-membrane lipid. This would mean that even a small transfer of lipid might have important physiological consequences. This hypothesis is borne out by the fact that there is a discrepancy between the total theoretical level of lipid transfer and the actual

Figure 2

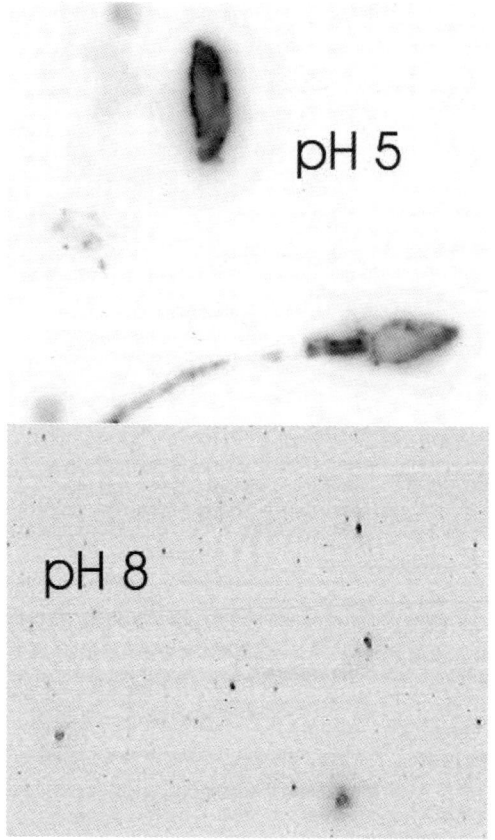

Fluorescence microscopic appearance of spermatozoa exposed to prostasomes labelled with R_{18} at pH 5.0 or 8.0 for 10 min

Magnification, ×520. The time of exposure was 3 s for pH 5 and 25 s for pH 8.

transfer obtained with the fusion process regardless of exposure time. Only 20% of the probe could be transferred at pH 5 in 10–15 min; this is in agreement with the hypothesis that only limited pools of spermatozoan lipids are affected by the fusion with prostasomes, and that acrosome-reacted sperm lose the ability to fuse.

However, polymorphonuclear and mononuclear leucocytes do not fuse to prostasomes under the same conditions as spermatozoa [23]. This fact hints at the specificity of prostasomal–spermatozoan fusion, although the fusion of prostasomes to other cell types has not been tested.

Transfer of protein from prostasome to sperm

By using R_{18}, we have demonstrated the transfer of lipid; however, prostasomes also possess some protein that is scarcely present in sperm plasma membrane and that may influence sperm function if transferred to spermatozoa [17]. We studied the transfer of CD13 (aminopeptidase N; EC 3.4.11.2) [24] and CD26 (dipeptidyl peptidase IV; EC 3.4.14.5) [25]. The results obtained with these two prostasomal peptidases were rather similar and confirm the results obtained previously with the R_{18} method. The advantage with using these enzymes is the possibility of both determining enzyme activity and following them using a specific, labelled antibody [24,25]. Spermatozoa acquire these activities in a pH-dependent manner when mixed with prostasomes. Cytofluorimetric determination confirmed the data obtained by measuring enzymic activity. Therefore, spermatozoa may acquire protein from prostasomes through the fusion mechanism. At the moment, it is impossible to state how many proteins will behave like CD13 and CD26. This mechanism may explain the changes observed in spermatozoa after ejaculation, even though protein synthesis is not active in these cells.

Effect of prostasome fusion on sperm cytosolic calcium

Calcium is present in low concentrations in the cytosol compared with the extracellular fluids. An impressive amount of literature deals with this phenomenon and it is well known that Ca^{2+} is involved in cellular signalling. Prostasomes are rich in calcium [26] and we tested the hypothesis that the sperm intracellular Ca^{2+} concentration ($[Ca^{2+}]_i$) may be affected by the fusion between spermatozoa and prostasomes. $[Ca^{2+}]_i$ was determined using the fura 2 method. The effect of mixing with prostasomes on sperm $[Ca^{2+}]_i$ depended on pH and on the prostasome/sperm ratio [27]. The effect on $[Ca^{2+}]_i$ was paralleled by the transfer of R_{18}. The $[Ca^{2+}]_i$ remained high until Na^+ was added to the external medium. This effect depended on an exchange between Ca^{2+} and Na^+, described in [28]. Therefore, *in vivo*, where Na^+ is present in the extracellular medium, fusion with prostasomes would produce a transient increase in sperm $[Ca^{2+}]_i$. This phenomenon may be discussed in light of two considerations: (i) lipid (cholesterol) transferred from prostasomes to sperm incapacitates the sperm [15,29,30], thus preventing an untimely acrosome reaction, and (ii) the increase in calcium increases sperm motility and induces the acrosomal reaction [31–33].

The treatment with 1 μM progesterone also increased spermatozoan $[Ca^{2+}]_i$, yet the treatment with progesterone of fused or non-fused spermatozoa showed that the effects of progesterone and prostasome fusion are additive phenomena (Figures 3 and 4). Experiments that were performed in the presence of external sodium showed that the increase in $[Ca^{2+}]_i$ due to fusion disappeared, but that the stimulating effect of fusion on the progesterone-induced increase in $[Ca^{2+}]_i$ did not. This means that the fusion mechanism modifies spermatozoa to make them more sensitive to the effect of progesterone, independently of its own action on $[Ca^{2+}]_i$.

Figure 3

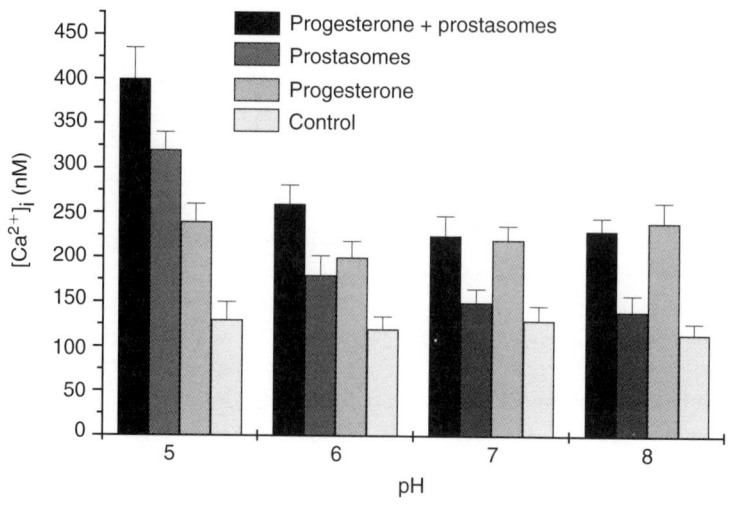

Effect of progesterone and fusion on spermatozoan [Ca²⁺]ᵢ in the absence of external Na⁺

Spermatozoa were mixed with prostasomes (prostasomal/spermatozoan protein ratio=2) for 15 min at 37 °C at the indicated pH values. Spermatozoa were collected and incubated for 10 min at 37 °C in sucrose/Hepes buffer without Na⁺. [Ca²⁺]ᵢ was then measured in either the presence or absence of 1 μM progesterone.

Figure 4

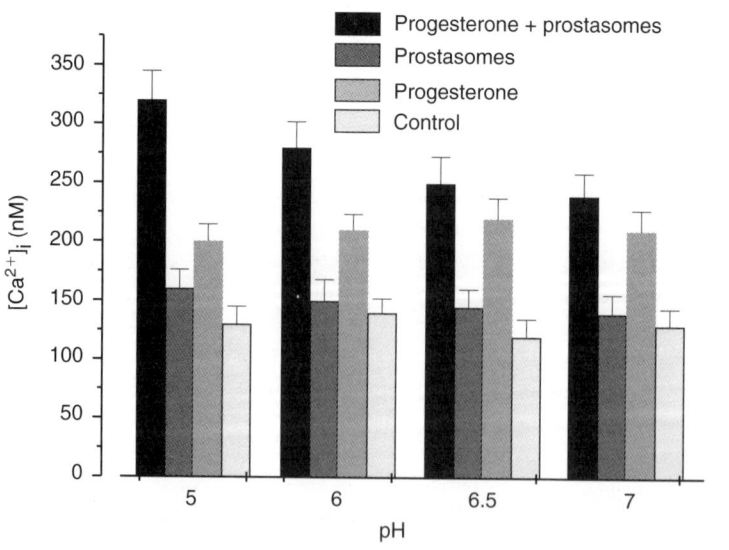

Effect of progesterone and fusion on spermatozoan [Ca²⁺]ᵢ in the presence of external Na⁺

For details, see Figure 3. The incubation medium contained 150 mM Na⁺.

	Acrosome-reacted sperm (%)						**Table 2**
	Control		+Ionophore		+Prostasomes	
	pH 5	pH 8	pH 5	pH 8	pH 5	pH 8
Not reacted	64±4	70±3	31±3	29±3	48±4	68±4
Partially reacted	12±2	8±1	24±2	28±2	22±2	10±1
Reacted	4±1	4±1	23±2	18±2	13±1	3±1
Broken	19±3	18±3	22±2	25±3	17±3	21±2

The effects of various treatments on the percentage of acrosome-reacted human sperm

Spermatozoa were exposed to pH 5 or 8 under various experimental conditions. The percentage of each cell type in each sample was calculated from a total of 200 cells.

Effect of prostasome fusion on the acrosome reaction

The acrosome reaction takes place in the proximity of the ovum, whereas fusion with prostasomes should occur in the vagina. The effect on $[Ca^{2+}]_i$ is transient under physiological conditions because of the exchange with Na^+ [28]. However, spermatozoa would be altered following the fusion with prostasomes and some alterations might last long enough to affect the acrosome reaction. The effects of progesterone on the acrosome reaction are mediated essentially by an increase in $[Ca^{2+}]_i$, stimulation of phospholipase activity, phosphorylation of protein and efflux of chloride [34]. The acrosomal reaction can be induced by the addition of progesterone or the calcium ionophore A23187 to sperm suspensions, and it can be assessed by microscopic observation after treatment with FITC-labelled *Pisum sativum* agglutinin. By this method, cells can be classified as reacted, not reacted, partially reacted and broken [35] (see Table 2). The exposure of spermatozoa to pH 5 did not affect the acrosomal reaction, either in controls or when the acrosomal reaction was triggered with the ionophore. Upon mixing of prostasomes and spermatozoa at pH 5, the number of acrosome reactions occurring was intermediate, between that of the controls and that obtained with the ionophore (Table 2).

Working hypotheses and conclusions

The fusion between prostasomes and spermatozoa may be one means of affecting spermatozoa reactivity and physiology. In our hands, fusion was pH-dependent and specific for spermatozoa. The lipid and protein patterns of spermatozoa changed upon fusion. Also, some biophysical parameters (e.g. fluidity) are modified by this phenomenon. In contrast, some modifications of spermatozoa produced by their contact with prostasomes do not appear to require fusion [36].

The physiological significance of fusion could be associated with the increased viscosity of spermatozoan membranes [13] and the increase in their cholesterol content. This lipid would decrease the possibility of untimely activation of spermatozoa. However, the fusion should stimulate spermatozoa that respond better to progesterone when this substance is used as an inducer of the acrosomal reaction. It should be noted, in connection with this, that spermatozoa are expected to lose cholesterol during their journey from the vagina to the fallopian tubes.

References

1   Ronquist, G. (1978) Eur. J. Clin. Invest. **17**, 213–236
2   Arienti, G., Carlini, E., De Cosmo, A.M., Di Profio, P. and Palmerini, C.A. (1998) Biol. Reprod. **59**, 309–313
3   Minelli, A., Moroni, M., Martinez, E., Mezzasoma, I. and Ronquist, G. (1998) J. Reprod. Fertil. **114**, 237–243
4   Stegmayr, B. and Ronquist, G. (1982) Urol. Res. **10**, 253–257
5   Kelly, R.W., Holland, P., Skibinski, G., Harrison, C., McMillan, L., Hargreave, T. and James, K. (1991) Clin. Exp. Immunol. **86**, 550–556
6   Skibinski, G., Kelly, R.W., Harkiss, D. and James, K. (1992) Am. J. Reprod. Immunol. **28**, 97–103
7   Lazarevic, M., Skibinski, G., Kelly, R.W. and James, K. (1995) Vet. Immunol. Immunopathol. **44**, 237–250
8   Lilja, H. and Laurel, C.B. (1984) Scand. J. Clin. Lab. Invest. **44**, 447–452
9   Carlsson, L., Pahlson, C., Bergquist, M., Ronquist, G. and Stridsberg, M. (2000) Prostate **44**, 279–286
10   Carlsson, L., Lennartsson, L., Nilsson, B.O., Nilsson, S. and Ronquist, G. (2000) Eur. Urol. **28**, 468–474
11   Arvidson, G., Ronquist, G., Wikander, G. and Ojteg, A.C. (1989) Biochim. Biophys. Acta **984**, 167–173
12   Arienti, G., Carlini, E., Polci, A., Cosmi, E.V. and Palmerini, C.A. (1998) Arch. Biochem. Biophys. **358**, 391–395
13   Carlini, E., Palmerini, C.A., Cosmi, E.V. and Arienti, G. (1997) Arch. Biochem. Biophys. **343**, 6–12
14   Benoff, S. (1993) Hum. Reprod. **8**, 2001–2008
15   Cross, N.L. and Mahasreshti, P. (1997) Arch. Androl. **39**, 39–44
16   Cross, N.L. and Razy-Faulkner, P. (1997) Biol. Reprod. **56**, 1169–1174
17   Fabiani, R. (1994) Upsala J. Med. Sci. **99**, 73–111
18   Hoekstra, D., de Boer, T., Klappe, K. and Wilschut, J. (1984) Biochemistry **23**, 5675–5681
19   Arienti, G., Carlini, E. and Palmerini, C.A. (1997) J. Membr. Biol. **155**, 89–94
20   Raboch, J. and Skaková, J. (1965) Fertil. Steril. **16**, 252–256
21   Wolf, D.E., Scott, B.K. and Millette, C.F. (1986) J. Cell Biol. **103**, 1745–1750
22   Wolf, D.E. (1995) Mol. Membr. Biol. **12**, 102–104
23   Arienti, G., Carlini, E., Saccardi, C. and Palmerini, C.A. (1998) Biochim. Biophys. Acta **1425**, 36–40
24   Arienti, G., Carlini, E., Verdacchi, R., Cosmi, E.V. and Palmerini, C.A. (1997) Biochim. Biophys. Acta **1336**, 533–538
25   Arienti, G., Polci, A., Carlini, E. and Palmerini, C.A. (1997) FEBS Lett. **410**, 343–346
26   Stegmayr, B., Berggren, P.O., Ronquist, G. and Hellman, B. (1982) Scand. J. Urol. Nephrol. **16**, 199–203
27   Palmerini, C.A., Carlini, E., Nicolucci, A. and Arienti, G. (1999) Cell Calcium **25**, 291–296
28   Bradley, M.P. and Forester, I.T. (1980) FEBS Lett. **121**, 15–18
29   Cross, N.L. (1996) Biol. Reprod. **54**, 138–145
30   Cross, N.L. (2000) Biol. Reprod. **63**, 1129–1134
31   Suarez, S.S. (1996) J. Androl. **17**, 331–335
32   Suarez, S.S. and Dai, X. (1995) Mol. Reprod. Dev. **42**, 325–333
33   Ahmad, K., Bracho, G.E., Wolf, D.P. and Tash, J.S. (1995) Arch. Androl. **35**, 187–208
34   Baldi, E., Luconi, M., Bonaccorsi, L., Maggi, M., Francavilla, S., Gabriele, A., Properzi, G. and Forti, G. (1999) Steroids **64**, 143–148
35   Mendoza, C., Carreras, A., Moos, J. and Tesarik, J. (1992) J. Reprod. Fertil. **95**, 755–763
36   Arienti, G., Carlini, E., Nicolucci, A., Cosmi, E.V., Santi, F. and Palmerini, C.A. (1999) Biol. Cell **91**, 51–54
37   Corazzi, L., Pistolesi, R. and Arienti, G. (1991) J. Neurochem. **56**, 207–212

15

Prostasomal effect on sperm motility

Lena Carlsson*[1], Anders Larsson*, B. Ove Nilsson† and Gunnar Ronquist*
*Department of Medical Sciences, Clinical Chemistry, University Hospital, SE-751 85 Uppsala, Sweden, and †Department of Medical Cell Biology, Unit of Anatomy, Biomedical Center, P.O. Box 571, SE-751 23 Uppsala, Sweden

Background

Sperm motility is probably one of the most important factors governing the fertilizing ability of human semen. In the lower female reproductive tract sperm motility is needed to penetrate cervical mucus, whereas in the upper tract vigorous beating of the sperm tail is necessary for penetration of the zona pellucida [1].

One functional effect of the prostasomes is to promote the progressive motility of human sperm [2]. Buffer-washed spermatozoa lose their ability to move forwards, but addition of prostasomes to such spermatozoa makes them motile again [3,4]. Hence, the addition of prostasomes at a protein concentration of 0.7–0.8 g/l to buffer-washed spermatozoa activated about 70% of the sperm cells, followed by a plateau, whereas albumin, a plasma protein known to stimulate sperm motility, had a less pronounced effect at a maximum concentration of 2–3 g/l. This was followed by a sharp decrease in motility, which was reduced to nothing at higher concentrations. Accordingly, prostasomes demonstrated a higher specific activity for spermatozoa than albumin. This discrepancy in behaviour between prostasomes and albumin was more pronounced in the low protein concentration range [3,4].

Motility-stimulating effects of prostasome inclusion in swim-up medium

When prostasomes were included in a conventional swim-up medium the yield of motile spermatozoa was increased by about 30%, suggestive of a beneficial effect of prostasomes to the fertilization medium at assisted reproduction [4]. Cryopreservation of spermatozoa generally results in a 40% decrease in motility and a 10–15% decrease in the pregnancy rate, compared with fresh semen [5]. The success of an assisted reproductive technology programme using cryopreserved spermatozoa would therefore be limited because of poor recovery of motile cells from post-thawed samples [6].

Subsequent experiments in our laboratory showed that prostasome supplementation of the swim-up medium also enhanced the recovery of motile, frozen–thawed spermatozoa [6]. Figure 1 illustrates the recovery of motile and

[1]To whom correspondence should be addressed (e-mail lena.carlsson@medsci.uu.se).

immotile spermatozoa after swim-up. The swim-up medium, Earle's balanced salt solution, was supplemented with albumin (1.25 mg/ml), prostasomes (1.0 mg/ml) or a combination of both. The combination of albumin and prostasomes was most efficient with respect to the yield of motile spermatozoa, with an increase of 50% compared with the medium containing only albumin. Also, replacing the albumin in Earle's balanced salt solution with prostasomes resulted in a 30% increase of motile cells compared with albumin alone. It should be noted that the presence of prostasomes in swim-up medium of thawed samples of cryopreserved spermatozoa always led to a significantly increased portion of immotile spermatozoa, resulting in a greater absolute amount of spermatozoa (Figure 1). The reason for this is not clear at present. One explanation could be the prostasomal induction of hyperactive spermatozoa, leading them into a hypermetabolic state and ending up in energy crisis and sperm necrosis.

Prostasome-like granules from the PC-3 prostate cancer cell line

We examined PC-3 cells derived from an *in vitro*-grown cell line of a human prostate cancer [7] with regard to their reactivity to a monoclonal antibody (mAb78) directed against seminal prostasomes [8]. There was a staining indicative of prostasome-like granules in the PC-3 cells [9]. Accordingly, it was of interest to characterize the functional properties of the prostasome-like granules (here denoted PC-3 prostasomes) from the PC-3 prostate cancer cell line and compare them with

Figure 1

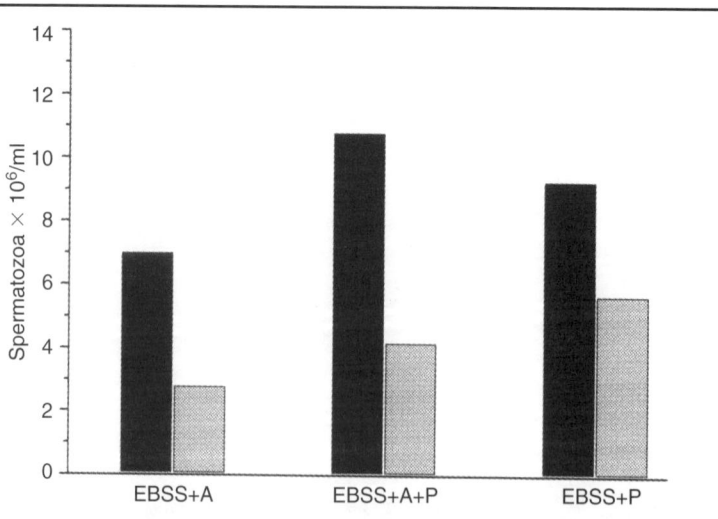

The numbers of motile and immotile spermatozoa recovered after swim-up

Shown are the numbers of motile (black bars) and immotile (grey bars) spermatozoa recovered after swim-up with Earle's balanced salt solution (EBSS) containing 1.25 mg/ml albumin (EBSS+A), 1.0 mg/ml prostasomes (EBSS+P) or both (EBSS+A+P).

those of the seminal prostasomes. We used computer-assisted sperm analysis (CASA) to investigate whether PC-3 prostasomes exhibited effects similar to those of seminal prostasomes on buffer-washed spermatozoa from normospermic semen samples. Further, by using immunostaining of prostasomes, we wanted to find out whether and where these granules were located on the spermatozoa.

Results

Motility-promoting effects of different concentrations of prostasomes and BSA

The effects of different concentrations of PC-3 prostasomes, seminal prostasomes and BSA on sperm motility were examined after 30 min of incubation (Figure 2). The proportion of motile sperm cells increased from about 40% with PC-3 prostasomes at a protein concentration of 0.05 mg/ml to about 55% at 0.1 mg/ml ($P<0.01$). Above this concentration, only minor changes were apparent. Neither was there any significant difference between PC-3 prostasomes and seminal prostasomes at any of the different concentrations tested, except at the highest

Figure 2

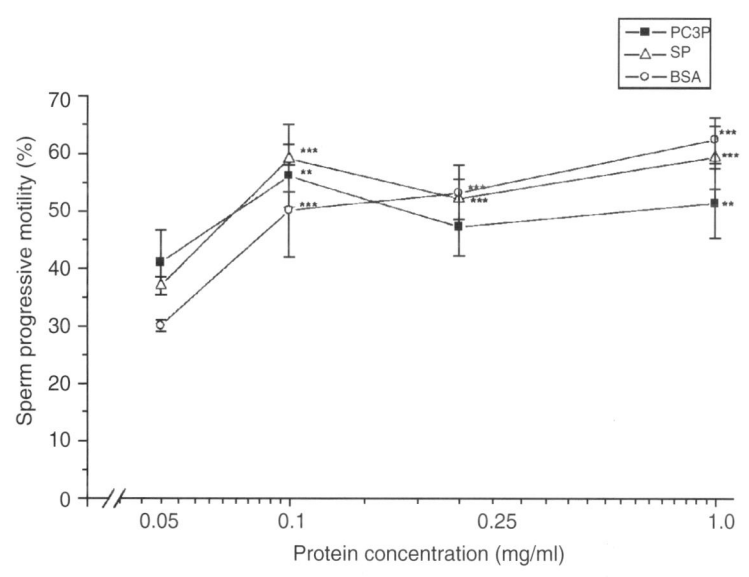

Effects of different concentrations of PC-3 prostasomes (PC3P), seminal prostasomes (SP) and BSA on sperm progressive motility after incubation for 30 min

*Prostasome and BSA concentrations are expressed as protein concentration (mg/ml). Results are means ±S.E.M. (n=7 in all groups). **P<0.01, ***P<0.001; n.s., not significant. All comparisons are made with the lowest protein concentration (0.05 mg/ml). The control value, i.e. the sperm progressive motility in buffer lacking any effectors, was 15±3% in all cases (results not shown).*

protein concentration (1.0 mg/ml; $P<0.05$). Significant increases in motility were found with BSA and seminal prostasomes at all three concentrations.

Comparison of motility-promoting effects between PC-3 prostasomes, seminal prostasomes and BSA over time

The effects of PC-3 prostasomes, seminal prostasomes and BSA on sperm cell motility over time were compared at a protein concentration of 0.1 mg/ml (Figure 3). All three effectors significantly stimulated sperm motility (defined as the proportion of motile cells). For instance, at 60 min of incubation, control sperm had a motility of about 15%, whereas prostasome- or albumin-supplemented sperm had a motility of about 50% ($P<0.001$). There were no significant differences between the two types of prostasome in terms of their stimulatory ability, and their efficiency decreased with time in a similar fashion.

Effect of heat treatment on the motility-promoting effect of PC-3 prostasomes, seminal prostasomes and BSA

Heat treatment did not significantly influence the stimulatory ability of PC-3 and seminal prostasomes at a protein concentration of 0.25 mg/ml, but heat treatment of BSA resulted in a decrease in sperm motility from about 50% to about 15% (i.e.

Figure 3

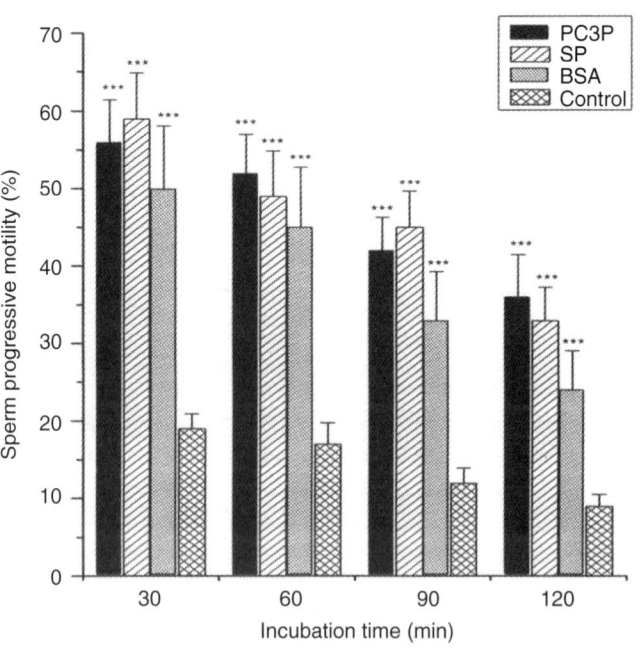

Effects of PC-3 prostasomes (PC3P), seminal prostasomes (SP) and BSA on sperm progressive motility after different incubation times

The protein concentration in each case is 0.1 mg/ml. Results are means ±S.E.M., n=7. ***P<0.001; all comparisons are made with the respective control (washed sperm without any effectors).

the control value) at 30 min of incubation ($P<0.001$; Figure 4). However, heat-treated prostasomes showed a decrease in activity over time similar to that of non-heated prostasomes. Discrepant behaviour was also observed between heat-treated albumin (heat-labile due to denaturation) and heat-treated prostasomes (heat-stable), indicating that the sperm-motility-promoting effect was not an isolated protein effect.

Immunocytochemical staining

Immunostaining with the monclonal antibody mAb78 resulted in a positive staining of all sperm cells coated with PC-3 prostasomes or seminal prostasomes. The staining, which occurred all over the spermatozoa, was intense on the mid-pieces and weaker on the sperm heads. Several sperm cells demonstrated a differential staining of the head, since the acrosome was more positive than the post-nuclear cap. The staining of the tails was weak.

Figure 4

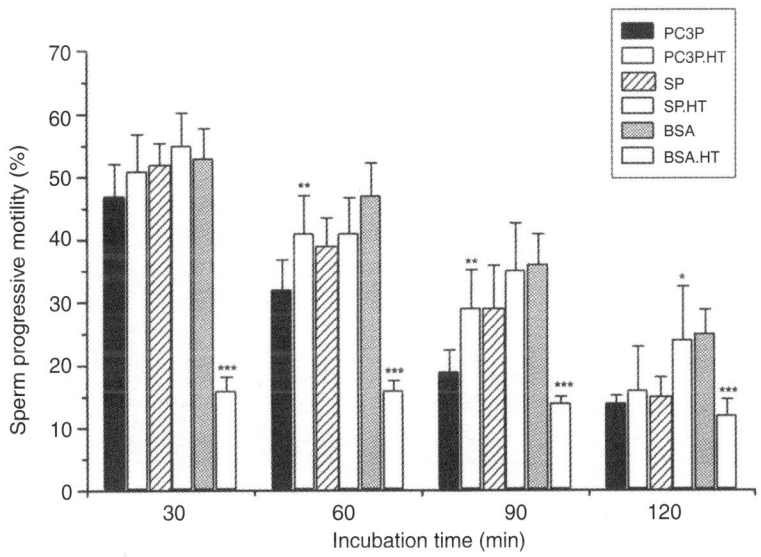

Effects of heat treatment (HT) of PC-3 prostasomes (PC3P), seminal prostasomes (SP) and BSA on sperm progressive motility after different incubation times

*The protein concentration in each case is 0.25 mg/ml. Results are means ±S.E.M., n=8. ***P<0.001; n.s., not significant. Comparisons are made in the respective pairs, i.e. no heat treatment versus heat treatment. The control value, i.e. the sperm progressive motility in buffer lacking any effectors, was 12±4% in all cases (results not shown).*

Discussion

Buffer washings of spermatozoa from normospermic semen samples resulted in a gradual loss of their forward motility, but the spermatozoa were functionally restored by the addition of seminal prostasomes [3,10]. Seminal prostasomes were also found to be useful for increasing the motility of frozen–thawed spermatozoa [6]. The present evaluation of the effect of the addition of PC-3 prostasomes on the number of motile spermatozoa was made on normospermic cells whose motility was reduced by repeated washings in an isotonic Tris/HCl buffer. We found that the addition of such prostasomes increased the proportion of motile spermatozoa from about 15% to 50–70%, which was satisfactory. The positive effect of the PC-3 prostasomes on the spermatozoa lasted only for a limited period of time, most probably because of a lack of energy-providing components, and generally there was little motility after 2 h. Thus the PC-3 prostasomes possess a motility-stimulating ability similar to that of seminal prostasomes. Different albumin preparations were investigated in a previous investigation with regard to their sperm-motility-promoting activity [3]. The comparison comprised both human and bovine albumin preparations, including one devoid of fatty acids. No major differences were found between the different preparations, and the BSA used in the present investigation was one of the albumin preparations studied previously [3].

Addition of albumin to the isotonic Tris/HCl buffer also increased the yield of motile sperm cells to an extent similar to that after addition of prostasomes. However, after moderate heat treatment, the motility-promoting effect of PC-3 prostasomes remained, whereas that of albumin diminished greatly, indicating different mechanisms of motility-promoting action.

Accordingly, the prostasome effect did not seem to be due to protein macromolecular structure. The seminal prostasomes also maintained their stimulatory ability after heat treatment, which agrees with previous findings [10]. A possible reason for the heat resistance of prostasomes may be the extraordinary architecture of the prostasome membrane. It has a highly ordered structure with an unusually high cholesterol/phospholipid ratio. This may account for the remarkable property of resistance against moderate heat treatment, which might include protection of an ordered protein structure [11].

The mechanism by which non-heated prostasomes initiate forward motility in buffer-washed spermatozoa is not known, nor is it known whether the mechanism requires a membrane contact or not [12]. However, since prostasomes were found to adhere to sperm cells when subjected to free-zone electrophoresis in a previous study [13] and when observed with immunocytochemistry in the present study, it seems that the working principle may include a close prostasome–sperm-cell contact. For instance, the properties of the sperm membrane, such as membrane permeability to Ca^{2+} and H^+, may be influenced. Other effectors in this context could be thermostable peptides, e.g. neuropeptide Y or vasoactive intestinal peptide, which are indeed components of the prostasomes [14]. They may modulate Ca^{2+} transport and stimulate sperm cAMP formation, thereby participating in the motility-promoting effect [15,16]. Further, we found previously by molecular biological methods that dipeptidyl peptidase

IV (also known as CD26) is a component of the prostasomes [17]. Since it has become apparent that dipeptidyl peptidase IV is able to mediate binding of cells to extracellular matrix proteins [18–21], we assume that this peptidase may play a role in the prostasome–sperm interaction. Our immunostaining study revealed that the whole sperm cell was stained and that the stain was most intense on the mid-piece. This indicated that the PC-3 prostasomes could adhere to the sperm cell. Since the mid-piece is the region where the mitochondria are located, it is possible that some prostasome component activates mitochondrial function, with increased sperm motility as a result. Since substance P, which regulates the activation of acetylcholine, can be cleaved by dipeptidyl peptidase IV, the prostasome-bound activity of dipeptidyl peptidase IV may influence the regulatory effect of acetylcholine on sperm motility [22]. The prostasome-associated activity of dipeptidyl peptidase IV might therefore represent the molecular link through which the prostasomes exert their promoting effect on sperm forward motility. It is known that prostasomes fuse with spermatozoa at slightly acidic pH values [23,24], and that this fusion means a transfer of lipid and protein to spermatozoan membranes [25]. Since our experiments were carried out at pH 7.6, a fusion process was less probable. Also, work by Arienti et al. [26] showed that the sperm-motility-promoting effect of prostasomes was not pH-dependent, ruling out fusion as an obligatory step preceding the stimulation of sperm motility.

Conclusion

We found that both PC-3 prostasomes and seminal prostasomes were able to reactivate washed and almost immotile sperm cells in the same fashion. Since the prostasomes adhered to the sperm cells, it seems that the ability to influence sperm cell motility also involves membrane-to-membrane contact. Considering that the spermatozoa were suspended in an isotonic Tris/HCl buffer without any supplementation, the results suggest that PC-3 prostasomes have a marked ability to restore different aspects of sperm cell motility.

References

1   Yanagimachi, R. (1966) J. Reprod. Fertil. **11**, 359–370
2   Stegmayr, B. and Ronquist, G. (1982) Urol. Res. **10**, 253–257
3   Fabiani, R., Johansson, L., Lundkvist, Ö., Ulmsten, U. and Ronquist, G. (1994) Eur. J. Obstet. Gynecol. Reprod. Biol. **57**, 181–188
4   Fabiani, R., Johansson, L., Lundkvist, Ö. and Ronquist, G. (1994) Hum. Reprod. **9**, 1485–1489
5   Alfredsson, J., Gudmundsson, S. and Snaedal, G. (1983) Obstet. Gynecol. Surv. **38**, 305–313
6   Carlsson, L., Ronquist, G., Stridsberg, M. and Johansson, L. (1997) Arch. Androl. **38**, 215–221
7   Kaighn, M.E., Narayan, K.S., Ohnuki, Y., Lechner, J.F. and Jones, L.W. (1979) Invest. Urol. **17**, 16–23
8   Nilsson, B.O., Jin, M., Einarsson, B., Persson, B.E. and Ronquist, G. (1988) Prostate **35**, 178–184
9   Nilsson, B.O., Lennartsson, L., Carlsson, L., Nilsson, S. and Ronquist, G. (1999) Upsala J. Med. Sci. **104**, 199–206
10 Fabiani, R., Johansson, L., Lundkvist, Ö. and Ronquist, G. (1995) Eur. J. Obstet. Gynecol. Reprod. Biol. **58**, 191–198

11   Arvidson, G., Ronquist, G., Wikander, G. and Öjteg, A.C. (1989) Biochim. Biophys. Acta **984**, 167–173

12   Cross, N.L. and Mahasreshti, P. (1997) Arch. Androl. **39**, 39–44

13   Ronquist, G., Nilsson, B.O. and Hjertén, S. (1990) Arch. Androl. **24**, 147–157

14   Stridsberg, M., Fabiani, R., Lukinius, A. and Ronquist, G. (1996) Prostate **29**, 287–295

15   Amann, R.P., Hay, S.R. and Hammerstedt, R.H. (1982) Biol. Reprod. **27**, 723–733

16   Garbers, D.L., Tubb, D.J. and Hyne, R.V. (1982) J. Biol. Chem. **257**, 8984–8990

17   Schrimpf, S.P., Hellman, U., Carlsson, L., Larsson, A., Ronquist, G. and Nilsson, B.O. (1999) Prostate **38**, 35–39

18   Hanski, C., Huhle, T. and Reutter, W. (1985) Biol. Chem. Hoppe-Seyler **366**, 1169–1176

19   Dang, N.H., Torimoto, Y., Schlossman, S.F. and Morimoto, C. (1990) J. Exp. Med. **172**, 649–654

20   Loster, K., Zeilinger, K., Schuppan, D. and Reutter, W. (1995) Biochem. Biophys. Res. Commun. **217**, 341–347

21   Minelli, A., Allegrucci, C., Mezzasoma, I., Ronquist, G., Lluis, C. and Franco, R. (1999) Biol. Reprod. **61**, 802–808

22   Dwivedi, C. and Long, N.J. (1989) Biochem. Med. Metab. Biol. **42**, 66–70

23   Arienti, G., Carlini, E. and Palmerini, C.A. (1996) J. Membr. Biol. **155**, 89–94

24   Arienti, G., Polci, A., Carlini, E. and Palmerini, C.A. (1997) FEBS Lett. **410**, 343–346

25   Carlini, E., Palmerini, A., Cosmi, E.V. and Arienti, G. (1997) Arch. Biochem. Biophys. **343**, 6–12

26   Arienti, G., Carlini, E., Nicolucci, A., Cosmi, E.V., Santi, F. and Palmerini, C.A. (1999) Biol. Cell **91**, 51–54

16

Prostasomes are antigens for naturally occurring anti-sperm antibodies

Cinzia Allegrucci*[1], Anders Larsson†, B. Ove Nilsson‡, Alba Minelli§ and Gunnar Ronquist†
*Division of Animal Physiology, School of Biosciences, University of Nottingham, Sutton Bonington Campus, Loughborough LE12 5RD, U.K., †Department of Medical Sciences, Clinical Chemistry, University Hospital, SE-751 85 Uppsala, Sweden, ‡Department of Medical Cell Biology, Unit of Anatomy, Biomedical Center, P.O. Box 571, SE-751 23 Uppsala, Sweden, and §Dipartimento di Scienze Biochimiche e Biotecnologie Molecolari, Sezione di Biochimica Cellulare, Via del Giochetto, 06123 Perugia, Italy

Introduction

Human reproductive failure may be a consequence of aberrant expression of immunological factors. Although the relative importance of immunological factors in human reproduction remains controversial, substantial evidence suggests that human leucocyte antigens (HLA), anti-sperm antibodies (ASA), integrins, the leukaemia inhibitory factor (LIF), cytokines, anti-phospholipid antibodies, endometrial adhesion factors, mucins (e.g. MUC1) and uterine natural killer cells contribute to reproductive failure [1].

ASA in men and women are important causes of infertility [2–5]. The role of naturally occurring ASA as a cause of infertility was recognized in 1954 [6]. Since these early observations a large number of studies have been directed to investigate the significance of ASA-dependent infertility, including the aetiology of ASA formation, the sites and mechanism of antibody action and possible treatments. However, since published studies have used various approaches for the recognition and treatment of immunological infertility, confusion has resulted concerning the prevalence of ASA as well as the development of accurate assays to screen ejaculates for sperm-bound antibodies. The role of ASA in infertility as well as the potential for their treatment is becoming better defined with the progress in assisted reproductive technologies.

The reported prevalence of ASA varies depending on the modality of the immunological screening. Circulating ASA range from 9 to 36% in infertile couples. In contrast, the prevalence of ASA in the general population is approx. 0–2% [7]. In addition, although women generally possess higher ASA titres than men, ASA are found more frequently in men [1]. Some clinical conditions associated with a high prevalence of ASA have been recognized. Among them are vasectomy [8], genito-urinary infections [9], testicular carcinoma [10], cryptorchidism [11], HIV [12] and varicocele [13]; it is also associated with homosexuality [14].

[1]To whom correspondence should be addressed (e-mail cinzia.allegrucci@nottingham.ac.uk).

In general, because spermatozoa are not produced in women and are not present in men until puberty, spermatozoal antigens are foreign to the immune system of both adult men and women. In men, autoimmunization to sperm antigens is normally blocked because of their sequestration in the testis by the blood–testis barrier and in the epididymis by tight junctions. T-suppressor lymphocytes (CD8+) may also play a role in the defence against autoimmunization to sperm antigens in the male.

The breakdown of sperm immune tolerance may depend on a decrease in the number or activity of T-suppressor cells, on a decrement of factors in seminal fluid that recruit T-suppressor cells, on altered sperm antigenicity that results in an inadequate suppression of immune responses, or on a breakdown of the blood–testis or blood–excurrent-ductal barrier, resulting in inoculation with sperm antigens [3].

In females, ASA production in the genital tract may occur if spermatozoa are exposed to immune responses from mechanical or chemical disruption of the mucosal layer. ASA formation may also be induced from spermatozoa in the peritoneal cavity after transtubal passage or from inflammation after genital infection [2].

Variable susceptibility may relate to a defect in the presence of or response to immunosuppressive factors in semen [4]. There are numerous physiological locations in which ASA may be detected, including male or female serum, semen, ovarian follicular fluid, vaginal or cervical secretions, or as antigenic epitopes bound directly to the sperm outer plasma membrane [7].

Antibodies in the serum belong predominantly to the IgG isotype, whereas those found in the seminal plasma are predominantly the IgA isotype. The main regions of the sperm to which IgA and IgG ASA are bound are the tail and mid-piece. The two Ig classes do not compete on equal terms for the binding sites on the sperm cell. Locally produced IgA antibodies reach the spermatozoa and occupy the binding sites before IgG is transudated from the prostate into the seminal compartment. When IgA occurs on sperm, IgG is usually also present and there are indications that both Ig classes are needed for subfertility effects on sperm [7].

ASA inhibit fertilization by several means. ASA can agglutinate spermatozoa by impairing sperm motility and sperm penetration through the cervical mucus and can interfere with several processes, such as the acrosome reaction, sperm–zona pellucida penetration, sperm–oocyte recognition and fusion, and pre-implantation embryonic development [15–21]. ASA on the sperm surface can also participate in complement-dependent, neutrophil-mediated sperm cytotoxicity. Most IgG and IgM ASA-positive sera fix homologous complement on human sperm cells by the classical pathway and contribute either to the phagocytosis of sperm cells by activated macrophages or to the complement-induced membrane-attack-complex action that leads to destruction of the sperm cell [22,23].

Different treatments of ASA-mediated infertility have been considered. They include corticosteroid therapy, intra-uterine insemination (IUI), *in vitro* fertilization (IVF) and intra-cytoplasmic sperm injection (ICSI). The rationale for the corticosteroid treatment is to reduce the production of ASA, thereby obtaining a proportion of antibody-free sperm sufficient for fertilization, but the efficacy of corticosteroid treatment has not been proven definitely. Since the most established interference of ASA with fertility is represented by the impairment of cervical mucus penetration by antibody-coated sperm, IUI has been used widely for the treatment of male immunological infertility. However, the IUI is an effective treatment for low or moderate sperm autoimmunization. Recourse to assisted reproductive technologies (IVF or ICSI) is mandatory when other less invasive approaches have failed. These methods may be the first choice in the case of a high degree of sperm autoimmunization. Different studies have suggested that if a high percentage of sperm are bound with IgA and IgG ASA, the fertilization rates in IVF might be decreased severely, necessitating the use of ICSI [2]. However, ICSI is an expensive and invasive therapy and therefore finding alternative means of overcoming the immunological infertility is appropriate.

Although many aspects of infertility due to naturally occurring ASA have been clarified, some controversies remain. A complete understanding of the mechanisms behind immunological infertility, as well as improved diagnosis and treatment, depends on knowledge of specific sperm antigens capable of eliciting the production of ASA.

It has been reported that there is a heterogeneity of antigens recognized by human ASA, which implies that there is considerable diversity in the anti-sperm Ig isotypes [2]. The sperm antigens recognized by ASA are generally characterized incompletely. The characterization of such sperm antigens may provide leads for improved diagnosis of antibody-mediated infertility and for identifying contraceptive vaccine candidates.

In this context, the role of prostasomes and their interactions with sperm cells are relevant. Prostasomes, organelles secreted by human prostatic acinar cells and expelled into the seminal plasma at ejaculation, adhere to and, to some extent, fuse with sperm cells [24]. It has been demonstrated that washed spermatozoa that are incubated with seminal prostasomes *in vitro* become coated with prostasomes [25].

The interaction/fusion between prostasomes and spermatozoa can modulate sperm functions. For instance, prostasomes promote the progressive motility of human sperm cells [26,27], improve the fertilization rate in assisted reproductive technologies [28] and modify the biochemical characteristics of ejaculated sperm cells [29,30].

Prostasomes can also influence the expression of immunomodulating antigens on the sperm surface [31]. Besides inhibiting mitogen-induced lymphocyte proliferation and phagocytic cell activity [32,33], prostasomes expose on their surface several glycosylphosphatidylinositol-anchored complement regulatory proteins, i.e. CD59, CD55, CD46 and CDw52 [34]. Because of the multiple roles of prostasomes and their close contact with spermatozoa, a contribution to the antigenicity of spermatozoa cannot be ruled out. Therefore, prostasomes could be candidate antigens for ASA. It should be kept in mind, however, that the prostasome membrane is composite and that two-dimensional electrophoresis has revealed at least 80 different protein entities [35]. Hence, the antigenicity of prostasomes is complex.

Our objective in the present study was to investigate whether prostasomes could be antigens for ASA found in the serum of infertile patients.

Method of study

We studied the reactivity of chicken anti-prostasome antibodies with sperm cells in an agglutination test. We also tested the reactivity of serum positive for ASA from 20 infertile patients with spermatozoa using flow cytometry and with purified prostasomes using ELISA.

Antibodies

Chicken anti-prostasome antibodies were produced by immunization of chickens with purified seminal prostasomes and purified by poly(ethylene glycol) precipitation. Chicken antibodies were used in this study as they have biochemical advantages over traditional mammalian antibodies, due to evolutionary differences [36]. They eliminate interference caused by the human complement system, rheumatoid factor and cellular Fc receptors. Also, chicken antibodies resemble human auto-antibodies in their reactivity [37].

Sera

Sera of patients investigated for infertility were collected and tested for the presence of ASA using the Tray agglutination test (cut-off point of 1:16; the minimum dilution able to produce agglutination) [38,39]. We selected 20 sera of ASA-positive patients at random (15 men and five women). Sera from six ASA-negative subjects (three women and three men) were collected and partly pooled (normal human serum; NHS), after individual Tray agglutination tests.

Semen

Semen samples were obtained by masturbation from fertile men during evaluation for IVF. Only ejaculates that were consistently negative for sperm agglutination and which met the requirements of the World Health Organization guidelines [40] were used in this study.

Agglutination

To evaluate the reactivity of chicken anti-prostasome antibodies with sperm cells, spermatozoa (2×10^6) were incubated with increasing concentrations of chicken polyclonal anti-prostasome antibody (0.08–3 mg/ml) for 30 min at 37°C. The sperm cells were then observed for agglutination by light microscopy.

Flow cytometry

For flow cytometry analysis, sperm cells were analysed using an EPICS XL™ flow cytometer (Coulter, Hialeah, FL, U.S.A.). Data processing from 10000 spermatozoa was carried out with XL software (Coulter). The mean background fluorescence value±2 S.D. from the results of IgG- and the complement component C3-binding experiments performed on ASA-negative sera was used as the cut-off value for positive results. Sperm cells incubated with buffer were used as a control for NHS. Negative controls were performed using an irrelevant chicken antibody (2 mg/ml).

ELISA

For ELISA, plates were coated with 4 μg of purified seminal prostasomes. The plates were blocked overnight at 4°C with coating buffer containing 3% BSA and incubated with 200 μl of serum samples (diluted 1:5 in PBS) for 2 h at 37°C. Biotinylated chicken anti-human IgG antibody (0.5 mg/ml) and streptavidin-alkaline phosphatase conjugate (0.1 mg/ml) were used. Alkaline phosphatase substrate solution was added and the absorbance was measured at 405 nm in an ELISA reader system (Spectra Max 250; Molecular Devices). The absorbance obtained with the pooled ASA-negative sera (NHS) was considered as the cut-off value for positive results. Negative controls were performed by omitting the antigen.

Statistical analysis

All data were analysed by Student's *t* test and least-squares linear-regression analysis. The significance of correlation was determined by Pearson's correlation coefficient.

Results and discussion

Sperm cell agglutination by chicken anti-prostasome antibodies

Swim-up spermatozoa immunostained for prostasomes with chicken anti-prostasome antibody (0.5 mg/ml) were found to be coated with prostasomes, most pronounced on the mid-piece (Figure 1). Polyclonal chicken anti-prostasome antibodies caused strong agglutination of sperm cells. Approx. 80% of spermatozoa were agglutinated. The agglutination displayed several types of sperm formation, mostly tail-to-tail contacts, but the type of interaction was dependent upon the concentration of the anti-prostasome antibody (Figure 2). An irrelevant chicken antibody at the same protein concentration was incubated in the same way with spermatozoa, but no agglutination was present. Inhibition of anti-prostasome antibody-mediated agglutination was investigated by reacting the chicken polyclonal anti-prostasome antibodies (0.5 mg/ml) with increasing concentrations of seminal prostasomes (0.05–1.0 mg/ml) for 30 min at 37°C. The antibodies were then incubated with sperm cells for 30 min at 37°C and agglutination was observed by light microscopy. When anti-prostasome antibodies were pre-incubated with high concentrations of prostasomes, no agglutination was observed during the subsequent contact with spermatozoa.

Flow cytometric analysis of IgG and C3 binding to sperm cells

Flow cytometric analysis of IgG and C3 binding to sperm cells induced by patient sera was performed by incubating human spermatozoa (5×10^6) with 50 μl each of the 20 ASA-positive and six ASA-negative sera for 60 min at 37°C in 5% CO_2/95% air. After incubation, the cells were washed and the binding of IgG was determined by incubating the sperm cells with FITC-conjugated chicken anti-human IgG antibody (2.3 mg/ml). The binding of complement component C3 was analysed by incubating the sperm cells with FITC-conjugated chicken anti-human C3 antibody (2.0 mg/ml). IgG antibodies against sperm cells were detected in all the patient sera. In the majority of cases (90%), the sera of the patients caused complement activation, measured by the deposition of C3 on the sperm

Figure 1

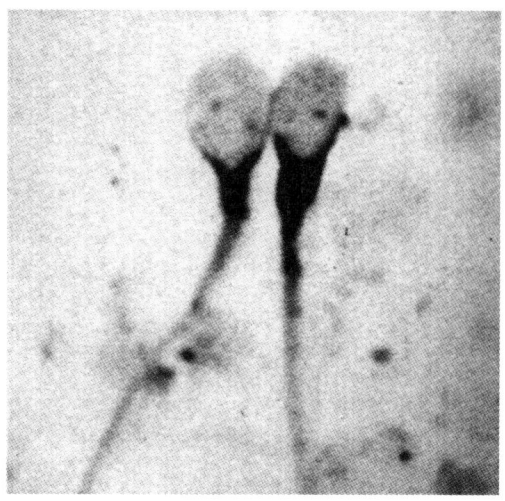

Immunostaining of swim-up sperm cells

Swim-up sperm cells were incubated with chicken anti-prostasome antibody (0.5 mg/ml) for 30 min at 37°C and with rabbit anti-chicken IgG biotin-conjugated antibody. The immunostaining demonstrates prostasomes coating the spermatozoa, especially on the mid-piece. Magnification, ×1800.

Figure 2

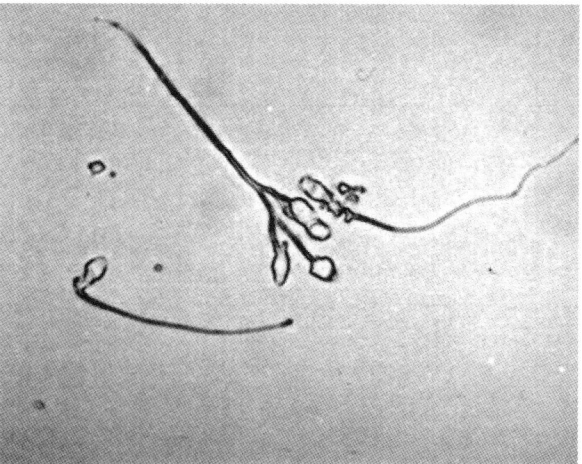

Agglutination of prostasome-coated sperm cells

Sperm cells (2 × 10⁶) were incubated with chicken anti-prostasome antibody (0.5 mg/ml) for 30 min at 37°C. Light micrograph of tail-to-tail agglutination, as seen in a routine preparation. Magnification, ×110.

cells (Figure 3). All sera used in these experiments had also been analysed for ASA titre by the Tray agglutination test. No significant correlation was found between the sperm-bound IgG and the ASA titre ($r=0.4$), whereas a significant positive correlation was found between sperm-bound C3 and the ASA titre ($r=0.43$) and especially between the expression of C3 and IgG ($r=0.79$).

ASA binding to prostasomes measured by ELISA

The 20 patient sera that were positive in the Tray agglutination test were analysed by ELISA for the presence of antibodies against prostasomes. Binding of IgG antibodies to prostasomes was clearly visible, although with variations in binding capacity (Figure 4). The absorbance values were significantly higher then those obtained with NHS as a negative control. No significant correlation was found between sperm cell agglutination capacity and the prostasome-binding affinity of each serum ($r=0.3$).

Conclusions

The present study shows unequivocally that human IgG-type ASA of infertile men and women recognize prostasomes as antigens. In addition, polyclonal antibodies raised against prostasomes are strongly immunogenic and they can at least contribute to immunological infertility. These results have interesting

Figure 3

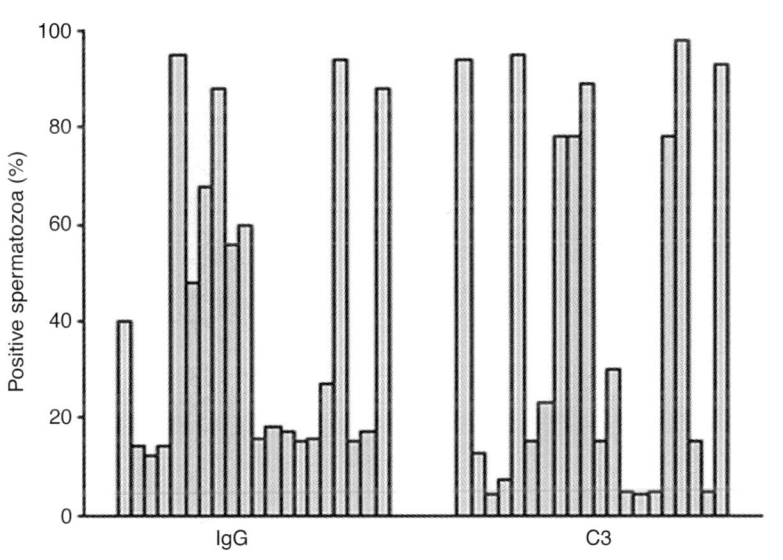

Flow cytometric determination of IgG and C3 binding to sperm cells

The mean background fluorescence ±2 S.D. from six ASA-negative sera was considered as the cut-off value for positive results. The results are expressed as the percentage of fluorescent spermatozoa.

Figure 4

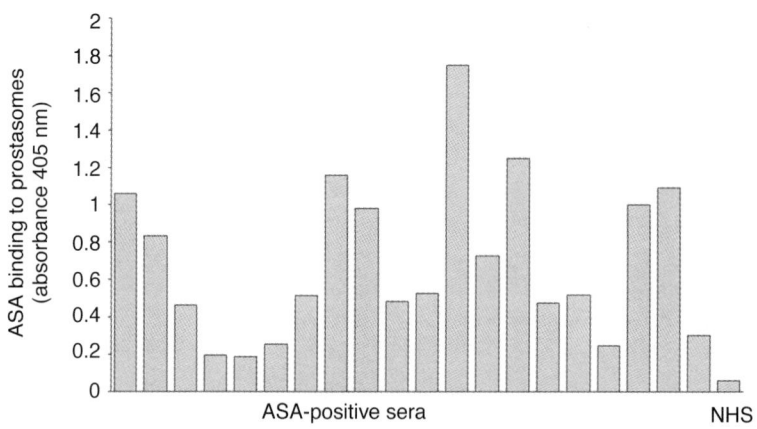

Binding of IgG-type ASA to prostasomes

ASA-positive sera were incubated with prostasome-coated ELISA plates. The plates were then incubated with anti-human IgG antibodies. Negative controls were performed by testing NHS.

implications for understanding of the mechanism involved in ASA-mediated infertility and for the immunological control of fertility.

The development of immunocontraceptive strategy has renewed interest in the study of clinical infertility mediated by ASA. ASA induced by immunization of men and women with antigens involved in natural immunoinfertility might similarly be without side effects.

Polyclonal antibodies are expected to interact with different antigens. Using hybridoma technology, specific monoclonal antibodies may become powerful tools for identification and distinction among human sperm antigens. Development of a vaccine based on sperm antigens is promising with respect to immunocontraception and offers an attractive approach to the growing global problem of overpopulation.

We acknowledge the contribution of Lena Carlsson and Monalill Lundqvist.

References

1   Choudhury, S.R. and Knapp, L.A. (2001) Hum. Reprod. Update **7**, 113–134
2   Mazumdar, S. and Levine, A.S. (1998) Fertil. Steril. **70**, 799–810
3   Meinertz, H. and Hjort, T. (1995) in Frontiers in Endocrinology, Immunocontraception (Nilsson, O. and Mattson R., eds), pp. 83–98, Ares-Serono Symposia Publications, Rome
4   Hjort, T. and Meinertz, H. (1995) in Frontiers in Endocrinology, Immunocontraception (Nilsson, O. and Mattson R., eds), pp. 151–159, Ares-Serono Symposia Publications, Rome
5   Marshburn, P.B. and Kutteh, W.H. (1994) Fertil. Steril. **61**, 799–811
6   Wilson, L. (1954) Proc. Soc. Exp. Biol. Med. **85**, 653–655
7   Heidenreich, A., Bonfig, R. Wilbert, D.M., Strohaier, W.L. and Engelmann, U.H. (1994) Am. J. Reprod. Immunol. **31**, 69–76
8   Broderick, G.A., Tom, R. and McClure, R.D. (1989) J. Urol. **142**, 752–755
9   Shahmanesh, M., Stedronska, J. and Hendy, W.F. (1986) Fertil. Steril. **46**, 308–311
10   Foster, R.S., Rubin, L.R., McNulty, A., Bihrle, R. and Donohue, J.P. (1991) Int. J. Androl. **14**, 179–185

11   Urry, R.L., Carrell, D.T., Strarr, N.T., Snow, B.W. and Middleton, R.G. (1985) J. Urol. **151**, 381–383
12   Naz, R.K., Ellaurie, M., Philips, T.M. and Hall, J. (1990) Biol. Reprod. **42**, 858–868
13   Knudson, G., Ross, L., Stuhldreher, D., Houliham, D., Bruns, E. and Prins, G. (1994) J. Urol. **151**, 1260–1262
14   Wolff, H. and Schill, W.B. (1985) Fertil. Steril. **44**, 673–677
15   Naz, R.K. and Menge, A.C. (1994) Fertil. Steril. **61**, 1001–1013
16   Mahony, M.C. and Alexander, N.J. (1991) Hum. Reprod. **6**, 1426–1430
17   Tasdemir, I., Tasdemir, M., Fukuda, J., Kodama, H., Matsiu, T. and Tanada, T. (1995) Int. J. Fertil. **40**, 192–195
18   Francavilla, F., Romano, R. and Santucci, R. (1991) Am. J. Reprod. Immunol. **25**, 77–80
19   Liu, Y. and Baker, H.W. (1992) Fertil. Steril. **58**, 465–483
20   Wolfe, J.P., De Almeida, M., Ducot, B., Rodriguez, D. and Jouannet, P. (1995) Fertil. Steril. **63**, 584–609
21   Naz, R.K. (1992) Biol. Reprod. **46**, 130–139
22   D' Cruz, O.J., Haas, Jr, G.G., Wang, B. and De Bault, L.E. (1991) J. Immunol. **146**, 611–620
23   Rooney, I.A., Oglesby, T.J. and Atkinson, J.P. (1993) Immunol. Res. **12**, 276–294
24   Ronquist, G., Nilsson, B.O. and Hjerten, S. (1990) Arch. Androl. **24**, 147–157
25   Wang, J., Lundqvist, M., Carlsson, L., Nilsson, B.O., Lundqvist, Ö. and Ronquist, G. (2001) Eur. J. Obstet. Gynecol. Reprod. Biol. **96**, 88–97
26   Stegmayr, B. and Ronquist, G. (1982) Urol. Res. **10**, 253–257
27   Fabiani, R., Johansson, L., Lundqvist, Ö. and Ronquist, G. (1995) Eur. J. Obstet. Gynecol. Reprod. Biol. **58**, 191–198
28   Carlsson, L., Ronquist, G. and Stridsberg, M. (1997) Arch. Androl. **38**, 215–221
29   Arienti, G., Carlini, E., Verdacchi, R., Cosmi, E.V. and Palmerini, C.A. (1997) Biochim. Biophys. Acta **1336**, 533–538
30   Arienti, G., Polci, A., Carlini, E. and Palmerini, C.A. (1997) FEBS Lett. **410**, 343–346
31   Kelly, R.W. (1999) Int. J. Androl. **22**, 2–12
32   Skibinski, G., Kelly, R.W., Harkiss, D. and James, K. (1992) Am. J. Reprod. Immunol. **28**, 97–103
33   Arienti, G., Carlini, E., Saccardi, C. and Palmerini, C.A. (1998) Biochim. Biophys. Acta **1425**, 36–40
34   Rooney, I.A., Heuser, J.E. and Atkinson, J.P. (1996) J. Clin. Invest. **97**, 1675–1686
35   Lindahl, M., Tagesson, C. and Ronquist, G. (1987) Urol. Int. **42**, 385–389
36   Larsson, A., Bålöw, R.M., Lindahl, T.L. and Forsberg, P.O. (1993) Poultry Sci. **72**, 1807–1812
37   Song, C.S., Yu, J.H., Bai, D.H., Hester, P.Y. and Kim, K.H. (1985) J. Immunol. **135**, 3354–3359
38   Friberg, J. (1974) Acta Obstet. Gynecol. Scand. **36**, 21–29
39   Androu, E., Mahmoud, A., Vermeulen, L., Schoonjans, F. and Comhaire, F. (1995) Hum. Reprod. **10**, 125–131
40   World Health Organisation (1999) WHO Laboratory Manual for the Examination of Human Semen and Sperm-Cervical Mucus Interaction, Cambridge University Press, Cambridge

17

Prostasome involvement in the acquisition of sperm fertilizing ability during epididymal transit

Robert Sullivan[1], Gilles Frenette and Christine Légaré

Unité d'Ontogénie-Reproduction, Centre de Recherche, Centre Hospitalier de l'Université Laval, 2705 Blvd. Laurier, Ste-Foy, PQ, Canada, G1V 4G2

Introduction: the epididymis and sperm maturation

The epididymis is present in all mammals, reptiles and teleostian fish. It is during transit through this long tubule that the spermatozoa acquire their ability to correctly encounter and fertilize the egg [1]. The epididymal lumen is composed of a pseudostratified epithelium that provides a microenvironment favourable to survival and maturation of the male gametes [2]. The composition of the intraluminal fluid varies depending on which section of the epididymis it is found in and what its content results from: absorption of most of the testicular fluid, selective transport of serum components or secretion of specific macromolecules. Certain interactions between the epididymal milieu and the sperm's surface must occur before fully functional male gametes can arise [3].

Hamster sperm P26h

Our team has previously identified a sperm surface protein, P26h, in the golden hamster *Mesocricetus auratus*. We proposed that it was involved in sperm–zona pellucida interactions, based on species-specific affinity of P26h for a homologous zona pellucida glycoprotein [4] and on inhibition of spermatozoa–zona pellucida binding by anti-P26h antibodies [5]. P26h is acquired by spermatozoa during epididymal transit and its localization is restricted to the membrane that covers the acrosomal cap. P26h, which shows evidence of being an integral membrane protein, can be removed from the surface of spermatozoa of the cauda epididymides using several different treatments. High-salt solutions are ineffective at removing P26h and treatment with a detergent is necessary. Because treatment with glycosylphosphatidylinositol (GPI)-specific phospholipase C releases P26h into the supernatant, we proposed that this sperm surface protein is anchored to the sperm plasma membrane via GPI [6]. GPI-anchoring of a protein originating from the extracellular milieu to the cell plasma membrane goes against the dogma that GPI-anchored proteins are first processed through the Golgi apparatus and secretory granules [7]. Knowing that proteins can be GPI-anchored to prostasomes

[1]To whom correspondence should be addressed (e-mail robert.sullivan@crchul.ulaval.ca).

and that proteins associated with prostasomes can be transferred to spermatozoa [8], we looked for prostasome-like particles (PLPs) in the epididymal lumen.

Yanagimachi et al. published a paper in 1985 [9] reporting that multi-lamellar vesicles are associated with spermatozoa in the caput-corpus epididymidis of Chinese hamsters, and they hypothesized that these vesicles could be involved in transferring cholesterol to the sperm plasma membrane to stabilize the membrane. Later, Fornes et al. [10] and Yeung et al. [8] reported the presence of PLPs in the epididymal fluid of rat, human and monkey.

In fact, P26h is found both associated with PLPs and in the soluble fraction (the supernatant produced by centrifugation at 100000 g) of epididymal fluid of sexually mature hamsters. PLPs are found throughout the length of the epididymis. Exposure of epididymal PLPs to phospholipase C results in the release of an important proportion of P26h into the supernatant [6]. Thus a significant proportion of P26h found in the epididymis is GPI-anchored to both spermatozoa and PLPs. These particles may be involved in the transfer of this protein to the sperm surface during epididymal maturation of the male gamete. This docking function of epididymal PLPs may also be involved in the addition of other proteins to spermatozoa transiting along the epididymis. In fact, a surprising number of proteins that are added to the sperm surface during epididymal maturation have been considered to be integral membrane proteins, based on physico-chemical properties [3]. To investigate the mechanisms of transfer of these proteins during gamete interactions, we looked for P26h orthologues in other mammalian species, including humans [11].

P34H, the human orthologue of hamster P26h

Using cDNA probes and specific antibodies, P26h orthologues have been fairly well characterized in mice [12], cattle [13], primates [14] and humans [15,16]. The human orthologue was named P34H, because of its electrophoretic mobility of 34 kDa. The P34H mRNA shares 67% sequence homology with hamster P26h and is expressed preferentially in the human corpus epididymidis [16]. During epididymal transit, P34H is added to the region of the sperm plasma membrane covering the acrosome [17]. Specific anti-P34H antibodies inhibit human sperm–zona pellucida binding *in vitro*, suggesting that P34H, like P26h in the hamster, is involved in this particular step of fertilization. Interestingly, very low levels of P34H are associated with spermatozoa in certain infertile men, whereas it is always present on spermatozoa of fertile men [18,19]. Thus suboptimal epididymal sperm maturation can be associated with some cases of infertility if we consider the presence of P34H on human spermatozoa as an indicator of epididymal maturation [11]. Due to certain obstacles that present when studying the human epididymis, the presence of PLPs has not yet been investigated adequately in humans, but we know that P34H is associated with PLPs in human semen (C. Légaré and R. Sullivan, unpublished work). To better understand the interaction between PLPs and spermatozoa in epididymal fluid and how this process can be defective in some infertile males, we set out to find the most suitable animal model available. A bovine model proved ideal for our purposes.

Bovine fertility: a well-documented commodity

The genetic superiority of the Canadian dairy Holstein is world-renowned, making the collection of semen from this dairy for artificial insemination an important industry in Canada. Bulls are selected according to the phenotype of F_1 cattle and bull fertility. The fertility data for each bull is based on the outcome of numerous artificial inseminations. These data are becoming very sophisticated and are adjusted by a linear statistical model to compensate for confounding effects such as age of the inseminated female, the month of insemination, the price of the bull's semen, and so on. These fertility data, expressed as the 60 to 90 days non-return rate (NRR), allow precise quantification and comparison of inter-individual male fertility [20]. Considering that these bulls are frequently slaughtered due to the F_1 phenotype or a decrease in the market demand for a particular bull's semen, fertility data, fresh and/or cryopreserved semen and, eventually, the epididymides become available. The large size of the bull's reproductive organs makes fluid recovery along the epididymis more convenient than for small laboratory animals. Bulls are slaughtered weekly, and the reproductive organs can be obtained rapidly at a very low cost.

The role of PLPs in the bovine epididymis

Epididymal and sperm physiology, including sperm–PLP interactions in the epididymis, can be correlated with fertility using the bovine model [11]. We wanted to answer following questions using bull reproductive tissues. (i) Are there sperm proteins related to P26h/P34H? (ii) If so, does the quantity of this protein associated with a constant number of spermatozoa vary between bulls and can it be related to fertility data? (iii) Can PLPs be implicated in acquisition of this protein by epididymal spermatozoa? Finally, (iv) can a defect in this process explain the inter-individual variability in the quantity of these epididymal proteins associated with spermatozoa?

Using polyclonal antibodies directed against purified hamster P26h, we looked for a homologous bull protein. In Western blots of ejaculated bull sperm protein extracts, a putative protein, P25b (for 25 kDa and bovine), was detected. Immunohistochemical studies revealed that this protein was located principally on the membrane covering the acrosome. When added to an *in vitro* fertilization system using bovine oocytes matured *in vitro*, the antisera partly inhibited fertilization, probably by interfering with sperm–zona pellucida binding (R. Sullivan, unpublished work).

P25b levels associated with a constant number of ejaculated spermatozoa have been investigated using semen samples collected from different bulls. P25b levels determined in different semen samples from one bull showed consistent values (with less than 10% difference between samples). In contrast, P25b showed high inter-individual variation. Low P25b values correlated with spermatozoa from bulls with a low NRR (i.e. low fertility rates), whereas high P25b values can be predicted from a high NRR (high fertility rates) [13]. Considering that fertility data are based on outcomes from thousands of artificially inseminated herds, the amount of P25b can give a good indication of the fertilizing ability of male gametes.

Epididymides from unmanipulated sexually mature bulls were obtained from the slaughterhouse immediately after slaughter to investigate whether P25b is associated with sperm maturation. Organs were kept on ice until dissection 3 h after slaughter. Only the epididymis with turgid distal tubules were used. Tubules from caput epididymis were sliced free with a surgical blade. Intraluminal fluid was collected by applying gentle pressure on the upper part of the dissected tubules. Cauda epididymal fluid was collected from dissected tubules following retrograde flushing under air pressure applied with a syringe within the scrotal section of the vas deferens. Using this procedure, 50–100 µl and 250–1500 µl of intraluminal fluid can be recovered from a pair of caput and cauda epididymides, respectively. Spermatozoa were pelleted by centrifugation at 300 g. Prostasomes were prepared by serial centrifugation. Epididymal fluids were first centrifuged at 700 g, the supernatant was then centrifuged at 3000 g and finally PLPs were pelleted by ultracentrifugation at 120 000 g. Following this same method, PLPs can be resuspended in an appropriate buffer for different experimental conditions or frozen at −80°C until use.

Western blotting of proteins extracted from spermatozoa recovered from different sections of the epididymis showed that caput spermatozoa have undetectable or low P25b levels; P25b is associated with spermatozoa collected more distally, being present in higher amounts in cauda sperm extract than in the corpus spermatozoa. It thus appears that this sperm surface protein accumulates on bull spermatozoa during epididymal transit, similar to related proteins in other species such as the hamster, mouse and human. The quantity of P25b associated with PLPs also shows great variability along the epididymis, with the quantity increasing during epididymal transit. Although the amount of P25b associated with PLPs collected in the caput region varies from undetectable to present in a significant amount, it is always much lower than in the cauda epididymides of the same animal. Being obtained from a slaughterhouse, the endocrine status of the animals from which the epididymides were collected remained unknown and could explain the inter-individual variability of P25b quantities associated with caput PLPs. We postulate that this bull-to-bull variation could be related to varying male fertility. This may explain, at least in part, the inter-individual variations in P25b quantities associated with a constant number of ejaculated spermatozoa of bulls characterized by different NRRs [21].

The anchoring mechanisms of P25b to cauda or ejaculated spermatozoa as well as to cauda epididymal PLP have been investigated. Hypertonic treatments of both spermatozoa and PLPs did not release P25b into the supernatant. This suggests that hydrostatic interactions are not involved in anchoring P25b to the sperm surface, as could be predicted for many other surface proteins acquired by spermatozoa during epididymal transit. When cauda epididymal or ejaculated spermatozoa are resuspended in an iso-osmotic solution and then treated with a phospholipase C specific for GPI, P25b is not released into the supernatant. Spermatozoa must be pretreated with 1 M NaCl to enable P25b release by this enzymic treatment. Thus hypertonic treatment probably removes coating proteins that mask P25b, which would be inaccessible to the enzyme when spermatozoa are prepared in iso-osmotic solutions. These proteins are probably released from the sperm surface during capacitation or transit into the female genital tract, or both, to expose P25b and enable fertilization. Similar treatments applied to cauda epididymal PLPs are unable to release P25b into the supernatant of ultracentifuged treated PLPs. Thus P25b is not exposed at the PLP surface and or masked by factors retained by hydrostatic interaction [21].

P25b is undetectable in the soluble fraction of epididymal fluid, is associated with epididymal PLPs and accumulates on the sperm surface during epididymal transit. Thus it can be hypothesized that PLPs may be involved in P25b transfer to the sperm surface, as has been shown for other proteins associated with prostasomes present in semen of different mammalian species. To investigate this possibility, PLPs prepared from the cauda epididymidal fluid were co-incubated with spermatozoa collected in the caput epididymides. Following incubation, the amount of P25b associated with a constant number of caput spermatozoa increased dramatically, as determined by Western blotting using specific antibodies. This increase of P25b associated with spermatozoa reached a 10-fold increase when compared with spermatozoa incubated without PLPs. Many mechanisms could be involved in this transfer: the integration of GPI-linked P25b with the sperm plasma membrane, very tight association of PLP with the sperm surface, or fusion/incorporation of these particles to the sperm plasma

Figure 1

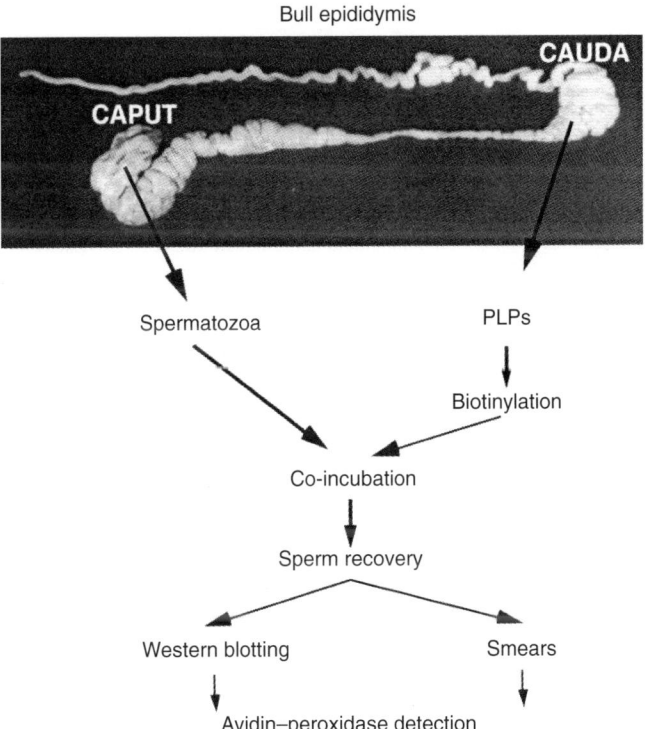

Procedure used to investigate interactions between cauda epididymal PLPs and caput spermatozoa in cattle

Figure 2

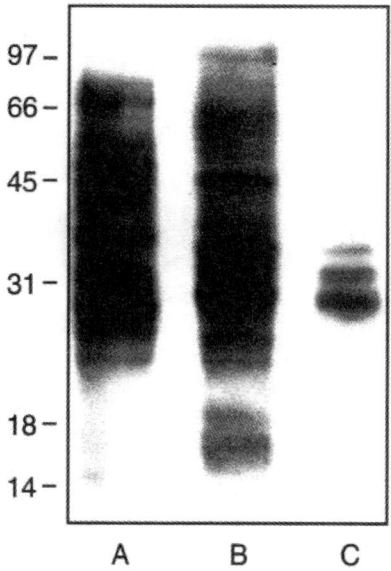

97 –
66 –
45 –
31 –
18 –
14 –

A   B   C

Western-blot analysis of biotinylated proteins

Biotinylated proteins of cauda spermatozoa (lane A), cauda epididymal PLPs (lane B) and transferred from cauda PLPs to caput spermatozoa (lane C) are shown. Proteins were revealed using peroxidase-conjugated avidin. Protein molecular-mass standards (in kDa) are indicated on the left.

membrane. These studies were done using antibodies that recognize P25b specifically. This protein may not be the only PLP-associated protein to be transferred to the sperm during epididymal transit. To investigate this, and to document the transfer mechanisms further, another experimental strategy was implemented.

PLPs purified from cauda epididymal fluid were biotinylated on surface amine residues of proteins associated with PLPs. This enables only proteins exposed at the PLP surface, and thus accessible to biotinylation, to be probed. These proteins can then be identified by methods similar to Western blotting, or localized on the sperm surface by histochemical methods using peroxidase-coupled avidin as a probe (Figure 1).

When analysed by Western blotting, biotinylated cauda PLPs reveal a complex electrophoretic pattern of a great variety of different-sized proteins. When incubated with caput spermatozoa, only a portion of these proteins is transferred to spermatozoa (Figure 2). This suggests that, under these conditions, PLPs are not transferred or fused as an intact entity to the spermatozoa. Only selected proteins became associated with spermatozoa, and these appeared to be the same ones in different experiments as well as with different PLP preparations. The location of these proteins' transfer to the sperm surface can be visualized by revealing smears of caput spermatozoa with peroxidase-conjugated avidin/biotin reagents. The distribution of PLP proteins shows heterogeneity from one sperm cell to another. Some remained unstained whereas the majority stained preferentially on the acrosome and mid-piece (Figure 3). Thus PLPs appear to transfer

Figure 3

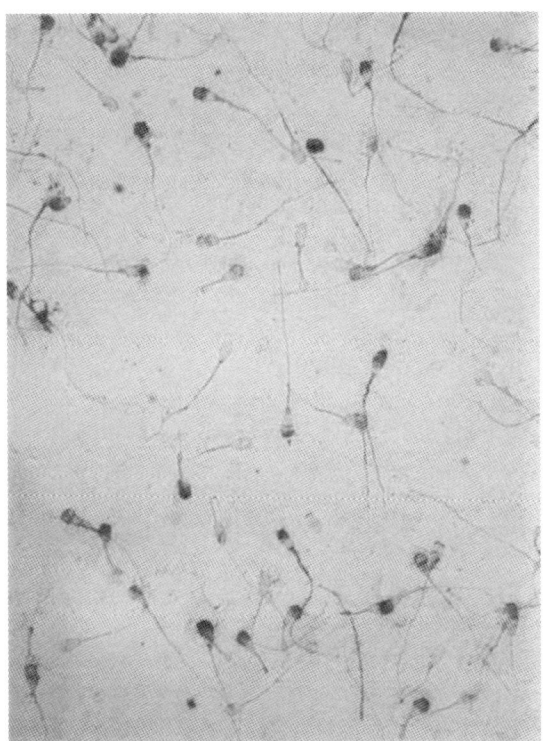

Histochemical localization of biotinylated proteins transferred from cauda epididymal PLPs to caput spermatozoa in cattle

Proteins were revealed using peroxidase-conjugated avidin.

selected proteins to specific sperm surface domains. If this is indeed the case, it is unlikely that PLPs fuse with the sperm plasma membrane. On the other hand, this may reflect prostasome heterogeneity. In this case, selected proteins transferred to sperm may be associated with only a subcategory of epididymal PLPs, which are the ones active in adding proteins to epididymal spermatozoa.

Even though Western-blot experiments revealed clearly that PLP-associated P25b is transferred to sperm during epididymal transit, P25b is not detected when transfer of biotinylated proteins from cauda PLP to caput spermatozoa is revealed with peroxidase-conjugated avidin. This is possibly due to the fact that this particular protein is not exposed at the PLP surface.

Transfer of biotinylated proteins from cauda PLPs to caput spermatozoa has been investigated under different experimental conditions. It has been reported by Carlini et al. [22] that fusion of sperm with prostasomes present in human semen occurs at a slightly acidic pH. The pH of human semen is at least 7.2, whereas bovine epididymal fluid has a pH of approx. 6.5 [23]. We therefore co-incubated cauda PLPs with caput spermatozoa at pH values between 5.5 and 7.5. The quantity of proteins transferred to spermatozoa was evaluated by densitometric determination of a Western blot of biotinylated transferred proteins probed with peroxidase-conjugated avidin. The transfer was less efficient at pH 7.5. The maximum transfer was reached at pH 6.0–6.5, with a 2.5-fold increase compared with the level reached at pH 7.5. At pH 5.5, approx. 20% fewer proteins were transferred on a constant number of spermatozoa compared with pH 6.0. At pH 6.5, which represents the physiological condition in the bull epididymis, the transfer appeared to be time- and temperature-dependent. At 32–37°C, the quantity of prostasomal proteins associated with spermatozoa doubled compared with the amount transferred at 22°C. At these temperatures, there was a progressive accumulation of protein with increased co-incubation time, until it reached a plateau after 120–150 min.

The effect of bivalent cations on prostasomal protein accumulation on caput spermatozoa was also investigated at pH 6.5. At concentrations of up to 2.5 mM, Mg^{2+} and Ca^{2+} had no significant effect on the rate of transfer. In contrast, Zn^{2+} had a beneficial effect. When compared with media not containing bivalent cations, the quantity of proteins transferred to a constant number of sperm increased with Zn^{2+} and almost doubled at a concentration of 1.25 mM.

Conclusions

During epididymal transit, proteins are transferred from the epididymal intraluminal milieu to the sperm surface. Some of these proteins behave as integral membrane proteins when exposed to different physico-chemical treatments. PLPs, or epididymosomes, are present in the epididymal fluid and are involved in transfer of such proteins to the maturing spermatozoa. This unusual mechanism of protein transfer appears to be involved in the acquisition of fertilizing ability of the male gamete during its trek along the male reproductive tract.

We thank Mr Carl Lessard for his help in preparing biological materials. This work was supported by a Natural Sciences and Engineering Research Council of Canada grant to R.S.

References

1   Jones, R.C. (1998) J. Reprod. Fertil. **53** (suppl.), 163–181
2   Hermo, L., Oko, R. and Morales, C.R. (1994) Int. Rev. Cytol. **154**, 105–189
3   Cooper, T.G. (1998) J. Reprod. Fertil. **53** (suppl.),119–136
4   Sullivan, R. and Bleau, G. (1985) Gamete Res. **12**, 101–116
5   Bérubé, B. and Sullivan, R. (1994) Biol. Reprod. **51**, 1255–1263
6   Légaré, C., Bérubé, B., Boué, F., Lefièvre, L., Morales, C., El Alfi, M. and Sullivan, R. (1999) Mol. Reprod. Dev. **52**, 225–233
7   Brown, D.A. and Rose, J.K. (1992) Cell **68**, 533–544
8   Yeung, C.-H., Schröter, S., Wagenfeld, A., Kirchhoff, C., Kliesch, S., Poser, D., Weinbauer, G.F., Nieschlag, E. and Cooper, T.G. (1997) Mol. Reprod. Dev. **48**, 267–275
9   Yanagimachi, R., Kamiguchi, Y., Mikamo, K., Suzuki, F. and Yanagimachi, H. (1985) Am. J. Anat. **172**, 317–330
10   Fornes, M.W., Sosa, M.A., Bertini, F. and Burgos, M.H. (1995) Andrologia **27**, 233–237
11   Sullivan, R. (1999) in The Male Gamete: From Basic to Clinical Applications (Gagnon, C., ed.), pp. 93–104, Cache River Press, Vienna, IL
12   Bégin, S., Bérubé, B., Boué, F. and Sullivan, R. (1995) Mol. Reprod. Dev. **41**, 249–256
13   Parent, S., Lefièvre, L., Brindle, Y. and Sullivan, R. (1999) Mol. Reprod. Dev. **52**, 57–65

14   Lamontagne, N., Légaré, C., Gaudreault, C. and Sullivan, R. (2001) Mol. Reprod. Dev. **59**, 431–441

15   Boué, F., Bérubé, B., De Lamirande, E., Gagnon, C. and Sullivan, R. (1994) Biol. Reprod. **51**, 577–587

16   Légaré, C., Gaudreault, C., St-Jacques, S. and Sullivan, R. (1999) Endocrinology **140**, 3318–3327

17   Boué, F., Blais, J. and Sullivan, R. (1996) Biol. Reprod. **54**, 1009–1017

18   Boué, F. and Sullivan, R. (1996) Biol. Reprod. **54**, 1018–1024

19   Guillemette, C., Thabet, M., Dompierre, L. and Sullivan, R. (1999) J. Androl. **20**, 214–219

20   Schaeffer, L.R. (1993) J. Dairy Sci. **76**, 837–842

21   Frenette, G. and Sullivan, R. (2001) Mol. Reprod. Dev. **59**, 115–121

22   Carlini, E., Palmerini, C.A., Cosmi, E.V. and Arienti, G. (1997) Arch. Biochem. Biophys. **343**, 6–12

23   Wales, R.G., Wallace, J.C. and White, I.G. (1966) J. Reprod. Fertil. **12**, 139–144

18

Protein kinases and protein phosphorylation in prostasomes

Michael J. Wilson

VA Medical Center, Minneapolis, MN, U.S.A., Department of Laboratory Medicine and Pathology, University of Minnesota, Minneapolis, MN, U.S.A, Department of Urologic Surgery, University of Minnesota, Minneapolis, MN, U.S.A., and Minnesota Cancer Center, University of Minnesota, Minneapolis, MN, U.S.A.

The post-translational modification of proteins by enzymic phosphorylation and dephosphorylation represents an important mechanism in regulating the biological operation of the affected molecules, and, in turn, a vast variety of cellular functions. Protein phosphorylation has been implicated in controlling metabolism, membrane transport, neurotransmission, genomic activation and cell proliferation [1–3]. Protein phosphorylation is carried out by protein kinases, a protein superfamily comprising approx. 2000 genes, which are related by virtue of their highly conserved catalytic domains, indicating they have evolved from a common primordial enzyme [4,5]. These enzymes can be subdivided into two main groups based on their ability to transfer the γ-phosphate of ATP (or GTP) to (i) alcohol groups on serine or threonine or (ii) phenolic groups on tyrosine residues of their protein substrates. The serine/threonine protein kinases can be categorized further on the functional basis of being either second-messenger-dependent [e.g. cAMP-dependent protein kinase (PKA), cGMP-dependent protein kinase, diacylglycerol-activated/phospholipid-dependent protein kinase C (PKC), Ca^{2+}/calmodulin-regulated protein kinase] or second-messenger-independent (e.g. protein kinase CK1 and protein kinase CK2). Protein kinases which have tyrosine as the target amino acid include receptor-associated kinases (e.g. insulin receptor and epidermal growth factor receptor) and oncogene products (e.g. Src, Ras and Abl). There are also protein kinases with dual specificity; that is, they can phosphorylate tyrosine and serine/threonine residues [e.g. the mitogen-activated protein kinases (MAPKs) extracellular-signal-regulated protein kinase (ERK) 1 and ERK2 can autophosphorylate at these three amino acids]. Proteins regulated by phosphorylation/dephosphorylation may be phosphorylated by more than one class of protein kinase and can have multiple phosphorylation sites. Protein phosphorylation is controlled reversibly by protein phosphatases, for which it is estimated that there are as many as 1000 genes [5]. Thus phosphorylation of target proteins is regulated by a complex network of protein kinases and phosphatases, widely distributed at the subcellular level, whose actions are modified by hormones, growth factors and mitogens.

Protein kinases and phosphatases are among the protein constituents secreted by the prostate gland. Although protein kinase activities had been found in sperm, Majumder [6] was first to report a protein kinase activity in seminal plasma. This protein kinase was cAMP-independent with a specificity for phosphorylating histones. Determination of enzymic molecular properties of protein kinase

Address correspondence to Research Service, VA Medical Center, One Veterans Drive, Minneapolis, MN 55417, U.S.A. (e-mail wilso042@tc.umn.edu).

activities towards the model basic and acidic protein phosphate acceptor-molecule substrates, lysine-rich histones and phosvitin, respectively, demonstrated the presence of multiple protein kinase activities in human seminal plasma [7]. Kinetic studies revealed more than one apparent K_m for ATP and the protein substrates. These protein kinase activities in common required Mg^{2+} and a thiol-protecting agent such as dithiothreitol, but the histone kinase activity was inhibited by increasing ionic strength and the cAMP-dependent protein kinase inhibitor protein (PKI), whereas phosvitin kinase activity was stimulated by increased ionic strength and was unaffected by the PKI. Neither protein kinase activity was affected by cAMP or cGMP. The inhibition by PKI and lack of cAMP stimulation of the histone kinase activity indicated that this kinase was the free catalytic subunit of PKA. Evaluation of protein kinase activities in seminal plasma from vasectomized men or those with normal sperm concentrations (>50 million/ml) demonstrated that the majority of these protein kinase activities in seminal plasma were from sperm, with approx. 20–30% originating from accessory sex glands [6,8]. Examination of protein kinase activities of split ejaculates from vasectomized men indicated that a significant portion of the accessory-sex-gland-derived histone kinase was of prostatic origin, whereas the phosvitin kinase was derived from both the prostate and seminal vesicles (Figure 1) [8]. The prostatic origin of histone and phosvitin kinase activities was substantiated by demonstration of these activities directly in prostatic secretions collected by prostatic massage [9].

The presence of protein kinase activities in the secretory granule and vesicle fraction of seminal plasma was first demonstrated by Stegmayr et al. [10].

Figure 1

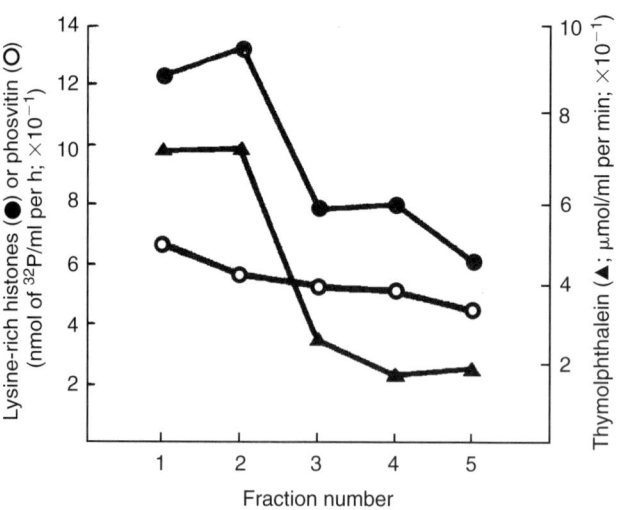

The phosvitin kinase (○), histone kinase (●) and prostatic acid phosphatase (▲) activities in sequential fractions of a split ejaculate from a vasectomized man

This distribution of protein kinase activities is representative of that observed in split ejaculates collected from five vasectomized men. Adapted from [8], with permission from the American Fertility Society for Reproductive Medicine.

Table 1

| Membrane fraction | PKI | Triton X-100 | Relative protein kinase activity | | |
			Phosvitin kinase	Histone kinase	Endogenous kinase
20 000 g	–	–	1.00	1.00	1.00
	+	–	1.13	0.13	1.19
	–	+	3.35	3.35	1.77
	+	+	3.40	0.34	1.77
105 000 g	–	–	1.00	1.00	1.00
	+	–	0.88	0.13	0.94
	–	+	5.01	4.96	3.51
	+	+	4.39	0.66	1.93

Effect of the cAMP-dependent PKI (400 ng/ml) and Triton X-100 (0.1%) on membrane-associated protein kinase activities of human seminal plasma from vasectomized men

Protein kinase reactions of these two membrane fractions of seminal plasma from vasectomized men were done in the presence or absence of 0.1% Triton X-100 and in the presence or absence of 400 ng/ml PKI. The data for each fraction are relative to the protein kinase activity in the untreated sample towards the individual protein substrate, i.e. phosvitin, lysine-rich histones and endogenous proteins. Data taken from [11].

Figure 2

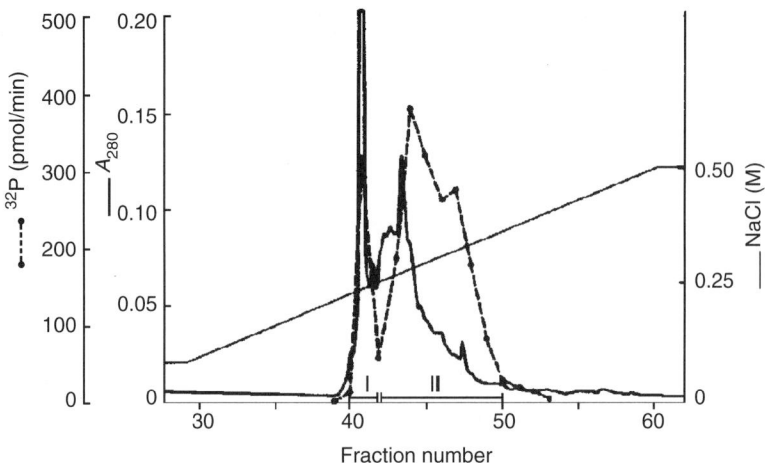

Chromatographic profile of membrane-associated histone kinase from membrane components of seminal plasma from a vasectomized man

They found that protein kinases in these organelles phosphorylated both serine and threonine residues in histones, phosvitin and endogenous organellar proteins. Phosphorylation of endogenous organellar proteins was not affected by cAMP, but cGMP stimulated threonine, but not serine, phosphorylation by 3.8-fold. There was no stimulation of phosvitin or histone phosphorylation by these cyclic nucleotides. The data from bivalent cations and different effector molecules indicated the presence of more than one kinase activity in prostasomes. We addressed prostasome protein kinase functions by preparing membrane fractions from seminal plasma of vasectomized men using differential centrifugation [11]. The membrane fractions in the post-cellular/nuclear supernatants accounted for 50% of the phosvitin and histone kinase activities and less than 7% of the seminal plasma protein. The phosphorylation of endogenous proteins was 3-fold higher in the membrane fractions as compared with the high-speed soluble fraction (Table 1). Phosphorylation of lysine-rich histones or phosvitin was not affected by cAMP or cGMP, but phosphorylation of endogenous membrane proteins was increased on exposure to these cyclic nucleotides by 55 and 71%, respectively. The protein kinase activities towards lysine-rich histones, but not phosvitin or endogenous proteins, were strongly inhibited by PKI, indicating that the histone kinase activity was due to the free catalytic subunit of PKA (Table 1). Differences in detergent extraction of the membranes further distinguished phosvitin and histone kinases, i.e. 80% of histone kinase and 13% of phosvitin kinase activities were extracted with lubrol. Anion-exchange chromatography on Mono Q separated the lubrol-extracted, membrane-associated histone kinase into two fractions (Figure 2), suggesting that two possible forms of histone kinase may exist in prostasomal membranes. Electrophoretic analysis showed a wide range of membrane proteins staining with a similar intensity from approx. 10 kDa to more than 120 kDa in size, whereas in the

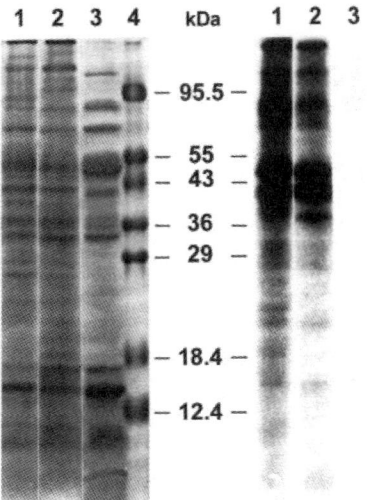

Figure 3

SDS/PAGE separation of endogenously phosphorylated membrane proteins from seminal plasma from vasectomized men

*Membrane suspensions of the 20 000 **g** (lanes 1) and 105 000 **g** (lanes 2) pellets and the 105 000 **g** supernatant (lanes 3) from the same vasectomy sample were phosphorylated and subjected to SDS/PAGE. The Coomassie Brilliant Blue-stained protein pattern is shown on the left and the autoradiograph on the right: 150 μg of protein were applied to each lane. Lane 4 shows the protein standards. Reproduced from [11], with permission from the Journals of Reproduction and Fertility, Ltd.*

Source or secretion type	n	CK2 activity (nmol of ^{32}P/ml per h)	**Table 2**
Normozoospermic	49	22.5±2.8	
Oligozoospermic	24	10.3±2.0*	
Azoospermic	6	8.1±2.1*†	
Vasectomized	38	4.9±0.4*	
Prostatic secretion	19	9.8±1.6	
Seminal vesicle secretion	5	5.5±1.7	

*P<0.01 compared with normozoospermic individuals.
†P<0.05 compared with vasectomized individuals.

Protein kinase CK2 activities measured with the CK2-specific synthetic peptide substrate RRREEETEEE in prostatic and seminal vesicle secretions and seminal plasmas of men who were vasectomized, azoospermic, oligozoospermic and or who had normal sperm concentrations (>20 million/ml)

Azoospermia was due to testicular failure; there was no obstruction of the vas deferens. Means ±S.E.M. are shown. Data taken from [12].

Figure 4

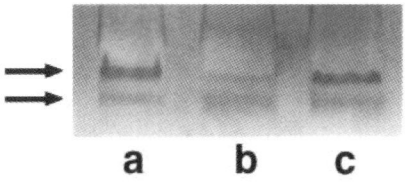

a   b   c

An immunoblot of protein kinase CK2 expression in human seminal plasma

Aliquots of seminal plasma from three non-vasectomized men (2 mg of protein) were semi-purified using phosphocellulose P-11, before separation by SDS/PAGE and transfer to a nitrocellulose sheet. The nitrocellulose sheet was probed with a purified polyclonal chicken antibody against rat liver CK2. The individual preparations of seminal plasma (specific CK2 activity of approx. 2.3 nmol [^{32}P]ATP/mg of protein per h) from men who had sperm concentrations of 150×10^6, 15×10^6 and 200×10^6 sperm/ml are represented in lanes a, b and c, respectively. A prominent band of the CK2 α-subunit is observed at 42 kDa and a less-intense band of α' is observed at 38 kDa (arrows). Reproduced from [12], with permission from the American Society of Andrology.

soluble fraction of seminal plasma from vasectomized men there were prominent proteins of approx. 65, 71–89 and 92 kDa (Figure 3). In contrast, there was strong phosphorylation of seven proteins of 38–120 kDa in the membrane fractions and only low-intensity phosphorylation of proteins in the soluble fraction.

Protein kinase CK2 has been identified as one of the protein kinases in human seminal plasma that phosphorylates acidic-type protein substrates, i.e. phosvitin, in human prostatic secretions. This was demonstrated by selective phosphorylation of the CK2-specific synthetic decapeptide substrate RRREEETEEE (Table 2) and immuno-identification with an anti-CK2 antibody (Figure 4) [12]. It remains to be established fully whether CK2 represents the phosvitin kinase localized in prostasomal membranes. However, CK2 is localized in several subcellular sites, including the plasma membrane [13], and phosvitin kinase activities in seminal plasma from vasectomized men are isolated predominantly in membrane fractions [11]. Despite the identification of CK2 and the free catalytic subunit of PKA in prostatic secretions and prostasomal membranes, other kinds of protein kinase have not been identified. There is an indication of Ca^{2+}-activated protein kinases in prostatic secretions; Ca^{2+} can partially substitute for Mg^{2+} in phosphorylation of lysine-rich histones and phosvitin [9]. Ca^{2+}-dependent protein kinases are stimulated by phospholipid or calmodulin, and these activators are in prostatic secretion [14,15]. An absence of tyrosine protein kinase in human seminal plasma has been reported [16].

The function of protein kinase activities in prostasomes is not known, but many of the factors that regulate protein kinase activities and the state of protein phosphorylation are present within human seminal plasma. The ATP concentration (approx. 60–130 μM) in human seminal plasma [17] is sufficient to support protein phosphorylation. cAMP is present in human seminal plasma and, based on the evaluation of split ejaculates, is most likely derived from the seminal vesicles [18]. PKI is in human seminal plasma and originates from accessory sex glands [19]. In addition, human prostatic acid phosphatase has phosphotyrosyl-protein [20,21] and phosphoserine/phosphothreonine protein phosphatase activities [22,23].

The role of protein kinases and the proteins that they may phosphorylate in prostasomes has been little studied. It is probable that they have some function in the secretory process itself. A number of prostasome proteins can be highly phosphorylated [11], and phosphorylated prostasome membranes are thicker than control prostasome membranes, as determined by electron microscopy [10]. In mammary epithelial cells, phosphorylation of casein molecules occurs in the Golgi, possibly mediated by two Ca^{2+}-dependent protein kinases, one calmodulin-dependent and the other calmodulin-independent. Phosphorylation of caseins is thought to help organize their packing in the secretory granules through cross-linking of the caseins with colloidal calcium phosphate [24]. Elevation of cAMP increases granule-to-granule fusion in pituitary lactotrophs [25] and increases granule movement in pancreatic β-cells [26]. Activation of PKC increases both granule-to-granule and granule-to-plasma-membrane fusion in lactotrophs [25] and a translocation of granules and myosin light-chain kinase to the pancreatic β-cell periphery; the latter effect is potentiated greatly by stimulation of both PKA and PKC [26].

Other roles of protein phosphorylation may be related to some of the functions attributed to prostasomes, such as sperm motility or immunomodulation. Observations indicating that prostatic secretions have an effect on sperm motility and survival were reported by Eliasson and Lindholmer [27]. The action of a "forward motility protein" in prostatic secretions on stimulation of sperm motility seemed evident [28], and this property appeared to reside in the prostasome fraction [29]. The increase in sperm motility may be due to the combined effects of molecular factors contributed by prostasomes, such as protein, ions, adenine or other constituents [30–33], and it may be due in part to protection of the spermatozoa against the acidic pH environment of the vagina [34]. A possibility is that sperm-surface protein kinases that phosphorylate sperm-surface proteins or proteins phosphorylated by sperm protein kinases [35] are derived from prostasomes through a transfer process similar to that for dipeptidyl peptidase IV [36] or aminopeptidase N [37]. Stegmayr et al. [10] noted a 10-fold increase in protein phosphorylation of washed sperm incubated with prostasomes compared with phosphorylation of each component alone. However, the motility-promoting effects of prostasomes are not prevented by heating, which suggests that they are not enzymic in nature [31].

Prostasomes appear to have an immunomodulatory function, possibly to protect sperm from complement attack in the female reproductive tract. They contain membrane cofactor protein CD46 [38], a type-I transmembrane protein providing complement component C3- and complement component C9-step inhibition, and the glycosylphosphatidylinositol-anchored proteins CD59 (inhibitor of the membrane-attack complex of the complement system), CD55 (decay-accelerating factor) and CDw52 [39,40]. Mechanisms of complement resistance in tumour cells include overexpression of these complement-regulatory proteins, but also protein phosphorylation, activation of G-proteins and turnover of phosphoinositides [41]. The protein phosphorylation mechanisms in tumour cells include phosphorylation of C9 by an ecto-protein kinase related to CK2 and activation of PKC and MAPK. Suppression of monocyte activation may involve interaction between CD46 and macrophage protein kinases that involve tyrosine phosphorylation on the cytoplasmic domain of CD46 [42].

There may be interactions of protein kinases and other prostasomal proteins. Dipeptidyl peptidase IV (CD26) is an ectopeptidase associated with prostasomes [43]. Inhibition of dipeptidyl peptidase IV catalytic function induces tyrosine phosphorylation (induces an inhibitory signal pathway), possibly p56 (lck), and p38 MAPK activation involved in the early phase of T-lymphocyte activation [44]. Neprilysin (CD10; neutral endopeptidase 24.11) is another prostasomal membrane peptidase. Neprilysin is phosphorylated in the cytoplasmic domain by CK2 and co-associates with additional tyrosine phosphoproteins including Lyn [45]. Granulophysin (CD63), a transmembrane 4 superfamily member, is associated with tyrosine protein phosphatases in leukaemia cells [46]. Tissue factor can be phosphorylated at multiple sites in its cytoplasmic domain [47]. Tissue factor has been known as a receptor for Factor VIIa involved in initiating the coagulation system, but this complex also generates intracellular signal transduction, resulting in activation of the p44/p42 MAPK pathway [48,49].

In summary, multiple protein kinase activities have been demonstrated to be associated with prostasomes. The roles of proteins phosphorylated by prostasomal protein kinases have not been established. However, protein phosphorylation has been implicated in functions attributed to prostasomes, such as sperm motility and immunomodulation, and phosphorylation of other proteins associated with the prostasomal membrane. Studies of protein phosphorylation in these systems should help clarify the role of prostasomes in semen.

I am extremely grateful to my long-time collaborators Dr Keith Kaye, Dr Akhouri Sinha, Dr Randolph Steer, Dr Jack Skalicky and Dr Khalil Ahmed in the studies reviewed here. Gratitude is also expressed for the funding support of this research by the Department of Veterans Affairs (Public Health Service grant CA 37718), awarded by the National Cancer Institute, Department of Health and Human Services, and the Minnesota Medical Foundation (grant CRF-48-81).

References

1   Edelman, A.M., Blumenthal, D.K. and Krebs, E.G. (1987) Annu. Rev. Biochem. **56**, 567–613
2   Hunter, T. (1991) Methods Enzymol. **200**, 3–37
3   Ahmed, K. (1994) Cell. Mol. Biol. Res. **40**, 1–11
4   Hanks, S.K. and Hunter, T. (1995) FASEB J. **9**, 576–596
5   Hunter, T. (1995) Cell **80**, 225–236
6   Majumder, G.C. (1978) Biochem. Biophys. Res. Commun. **81**, 1217–1226
7   Wilson, M.J., Steer, R.C. and Kaye, K.W. (1982) Biochim. Biophys. Acta **700**, 206–212
8   Wilson, M.J. and Kaye, K.W. (1983) Fertil. Steril. **40**, 105–109
9   Wilson, M.J., Steer, R.C. and Kaye, K.W. (1984) J. Lab. Clin. Med. **103**, 905–911
10   Stegmayr, B., Brody, I. and Ronquist, G. (1982) J. Ultrastruct. Res. **78**, 206–214
11   Wilson, M.J., Kaye, K.W., Skalicky, J. and Wilson, M.J. (1988) J. Reprod. Fertil. **83**, 635–646
12   Wilson, M.J., Davis, A., Ercole, C., Pryor, J.L., Hensleigh, H., Kaye, K.W., Dawkins, H.J.S., Wasserman, N.F., Reddy, P. and Ahmed, K. (1998) J. Androl. **19**, 754–760
13   Faust, M. and Montenarh, M. (2000) Cell Tissue Res. **301**, 329–340
14   Mann, T. and Lutwack-Mann, C. (1981) Male Reproductive Function and Semen, p. 307, Springer-Verlag, Berlin
15   Forrester, I.T. and Bradley, M.P. (1980) Biochem. Biophys. Res. Commun. **92**, 994–1001
16   Durocher, Y., Chapdelaine, A. and Chevalier, S. (1998) Cancer Res. **49**, 4818–4823
17   Singer, R., Barnet, M., Sagiv, M., Allalouf, D., Landau, B. and Servadio, C. (1983) Int. J. Biochem. **15**, 105–109
18   Stegmayr, B. and Ronquist, G. (1982) Scand. J. Urol. Nephrol. **16**, 91–95

19   Pliego, J.F., Van-Arsdalen, K. and Kopf, G.S. (1986) Biol. Reprod. **34**, 885–893

20   Li, H.-C., Chernoff, J., Chen, L.B. and Kirschonbaum, A. (1984) Eur. J. Biochem. **138**, 45–51

21   Lin, M.-F. and Clinton, G.M. (1986) Biochem. J. **235**, 351–357

22   Wasylewska, E., Czubak, J. and Ostrowski, W.S. (1983) Acta Biochim. Pol. **30**, 175–184

23   Lee, H., Chu, T.M. and Lee, C.L. (1991) Prostate **19**, 251–263

24   Burgoyne, R.D. and Duncan, J.S. (1998) J. Mammary Gland Biol. Neoplasia **3**, 275–286

25   Cochilla, A.J., Angleson, J.K. and Betz, W.J. (2000) J. Cell Biol. **150**, 839–848

26   Yu, W., Niwa, T., Fukasawa, T., Hidaka, H., Senda, T., Sasaki, Y. and Niki, I. (2000) Diabetes **49**, 945–952

27   Eliasson, R. and Lindholmer, C.H. (1972) Fertil. Steril. **23**, 252–256

28   Lindholmer, C.H. (1974) Biol. Reprod. **10**, 533–542

29   Stegmayr, B. and Ronquist, G. (1982) Scand. J. Urol. Nephrol. **16**, 85–90

30   Stegmayr, B. and Ronquist, G. (1982) Urol. Res. **10**, 253–257

31   Fabiani, R., Johansson, L., Lundkvist, O. and Ronquist, G. (1994) Hum. Reprod. **9**, 1485–1489

32   Fabrini, R., Johansson, L., Lundkvist, O. and Ronquist, G. (1995) Eur. J. Obstet. Gynecol. Reprod. Biol. **58**, 191–198

33   Carlsson, L., Ronquist, G., Stridsberg, M. and Johansson, L. (1997) Arch. Androl. **38**, 215–221

34   Arienti, G., Carlini, E., Nicolucci, A., Cosmi, E.V., Santi, F. and Palmerini, C.A. (1999) Biol. Cell **91**, 51–54

35   Miceli, D.C., Llanos, R., Jimenez, M. and Peralta, L. (1998) Zygote **6**, 203–212

36   Arienti, G., Polci, A., Carlini, E. and Palmerini, C.A. (1997) FEBS Lett. **410**, 343–346

37   Arienti, G., Carlini, E., Verdacchi, R., Cosmi, E.V. and Palmerini, C.A. (1997) Biochim. Biophys. Acta **1336**, 533–538

38   Kitamura, M., Namiki, M., Matsumiya, K., Tanaka, K., Matsumoto, M., Hara, T., Kiyohara, H., Okabe, M., Okuyama, A. and Seya, T. (1995) Immunology **84**, 626–632

39   Rooney, I.A., Atkinson, J.P., Krul, E.S., Schonfeld, G., Polakoski, K., Staffitz, J.D. and Morgan, B.P. (1993) J. Exp. Med. **177**, 1409–1420

40   Rooney, I.A., Heuser, J.E. and Atkinson, J.P. (1996) J. Clin. Invest. **97**, 1675–1686

41   Jurianz, K., Ziegler, S., Garcia-Schuler, H., Kraus, S., Bohana-Kashtan, O., Fishelson, Z. and Kirschfink, M. (1999) Mol. Immunol. **36**, 929–939

42   Wong, T.C., Yant, S., Harder, B.J., Korte-Sarfaty, J. and Hirano, A. (1997) J. Leukocyte Biol. **62**, 892–900

43   Vanhoof, G., De Meester, I., van Sande, M., Scharpe, S. and Yaron, A. (1992) Eur. J. Clin. Chem. **30**, 333–338

44   Kahne, T., Reinhold, D., Neubert, K., Born, I., Faust, J. and Ansorge, S. (2000) Adv. Exp. Med. Biol. **477**, 131–137

45   Ganju, R.K., Shpektor, R.G., Brenner, D.G. and Shipp, M.A. (1996) Blood **88**, 4159–4165

46   Carmo, A.M. and Wright, M.D. (1995) Eur. J. Immunol. **25**, 2090–2095

47   Mody, R.S. and Carson, S.D. (1997) Biochemistry **36**, 7869–7875

48   Ott, I., Fischer, E.G., Miyagi, Y., Mueller, B.M. and Ruf, W. (1998) J. Cell Biol. **140**, 1241–1253

49   Poulson, L.K., Jacobsen, N., Sorensen, B.B., Bergenhem, N.C., Kelly, J.D., Foster, D.C., Thastrup, O., Ezban, M. and Petersen, L.C. (1998) J. Biol. Chem. **273**, 6228–6232